Linear Algebra Demystified

Demystified Series

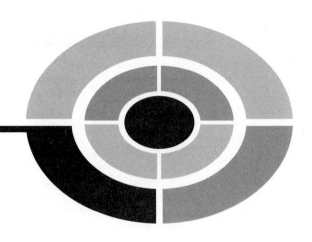

Linear Algebra Demystified

DAVID McMAHON

McGRAW-HILL

New York   Chicago   San Francisco   Lisbon   London   Madrid
Mexico City   Milan   New Delhi   San Juan   Seoul
Singapore   Sydney   Toronto

The McGraw·Hill Companies

Library of Congress Cataloging-in-Publication Data

McMahon, David.
    Linear algebra demystified / David McMahon.
        p.   cm.—(Demystified series)
    Includes bibliographical references and index.
    ISBN 0-07-146579-0 (acid-free paper)
    1. Algebras, Linear—Popular works.   2. Algebras, Linear—Problems, exercises, etc.
  I. Title. II. McGraw-Hill "Demystified" series.

QA184.2.M36   2006
512′.5—dc22   2005052282

2 3 4 5 6 7 8 9 0 DOC/DOC 0 1 0 9 8 7 6

ISBN 0-07-146579-0

The sponsoring editor for this book was Judy Bass and the production supervisor was Pamela A. Pelton. It was set in Times Roman by TechBooks. The art director for the cover was Margaret Webster-Shapiro.

Printed and bound by RR Donnelley.

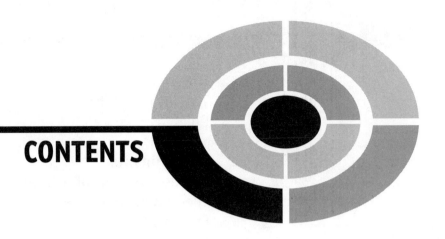

CONTENTS

CONTENTS

PREFACE

This book is for people who want to get a head start and learn the basic concepts of linear algebra. Suitable for self-study or as a reference that puts solving problems within easy reach, this book can be used by students or professionals looking for a quick refresher. If you're looking for a simplified presentation with explicitly solved problems for self-study, this book will help you. If you're a student taking linear algebra and need an informative aid to keep you ahead of the game, this book is the perfect supplement to the classroom.

The topics covered fit those usually taught in a one-semester undergraduate course, but the book is also useful to graduate students as a quick refresher. The book can serve as a good jumping off point for students to read before taking a course. The presentation is informal and the emphasis is on showing students *how* to solve problems that are similar to those they are likely to encounter in homework and examinations. Enhanced detail is used to uncover techniques used to solve problems rather than leaving the how and why of homework solutions a secret.

While linear algebra begins with the solution of systems of linear equations, it quickly jumps off into abstract topics like vector spaces, linear transformations, determinants, and solving eigenvector problems. Many students have a hard time struggling through these topics. If you are having a hard time getting through your courses because you don't know how to solve problems, this book should help you make progress.

As part of a self-study course, this book is a good place to get a first exposure to the subject or it is a good refresher if you've been out of school for a long time. After reading and doing the exercises in this book it will be much easier for you to tackle standard linear algebra textbooks or to move on to a more advanced treatment.

The organization of the book is as follows. We begin with a discussion of solution techniques for solving linear systems of equations. After introducing the

notion of matrices, we illustrate basic matrix algebra operations and techniques such as finding the transpose of a matrix or computing the trace. Next we study determinants, vectors, and vector spaces. This is followed by the study of linear transformations. We then devote some time showing how to find the eigenvalues and eigenvectors of a matrix. This is followed by a chapter that discusses several special types of matrices that are important. This includes symmetric, Hermitian, orthogonal, and unitary matrices. We finish the book with a review of matrix decompositions, specifically LU, SVD, and QR decompositions.

Each chapter has several examples that are solved in detail. The idea is to remove the mystery and show the student *how* to solve problems. Exercises at the end of each chapter have been designed to correspond to the solved problems in the text so that the student can reinforce ideas learned while reading the chapter. A final exam, with similar questions, at the end of the book gives the student a chance to reinforce these notions after completing the text.

David McMahon

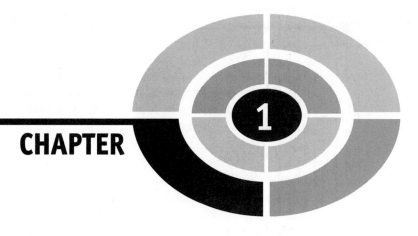

CHAPTER 1

Systems of Linear Equations

A *linear* equation with n unknowns is an equation of the type

$$a_1 x_1 + a_2 x_2 + \cdots + a_n x_n = b$$

In many situations, we are presented with m linear equations in n unknowns. Such a set is known as a *system of linear equations* and takes the form

$$a_{11} x_1 + a_{12} x_2 + \cdots + a_{1n} x_n = b_1$$
$$a_{21} x_1 + a_{22} x_2 + \cdots + a_{2n} x_n = b_2$$
$$\vdots$$
$$a_{m1} x_1 + a_{m2} x_2 + \cdots + a_{mn} x_n = b_m$$

The terms $x_1, x_2, \ldots, x_n$ are the *unknowns* or variables of the system, while the a_{ij} are called *coefficients*. The b_i on the right-hand side are fixed numbers or *scalars*. The goal is to find the values of the $x_1, x_2, \ldots, x_n$ such that the equations are satisfied.

EXAMPLE 1-1
Consider the system

$$3x + 2y - z = 7$$
$$4x + 9y = 2$$
$$x + 5y - 3z = 0$$

Does $(x, y, z) = (2, 1, 1)$ solve the system? What about $\left(\frac{11}{4}, -1, -\frac{3}{4}\right)$?

SOLUTION 1-1
We substitute the values of (x, y, z) into each equation. Trying $(x, y, z) = (2, 1, 1)$ in the first equation, we obtain

$$3(2) + 2(1) - 1 = 6 + 2 - 1 = 7$$

and so the first equation is satisfied. Using the substitution in the second equation, we find

$$4(2) + 9(1) = 8 + 9 = 17 \neq 2$$

The second equation is not satisfied; therefore, $(x, y, z) = (2, 1, 1)$ cannot be a solution to this system of equations.

Now we try the second set of numbers $\left(\frac{11}{4}, -1, -\frac{3}{4}\right)$. Substitution in the first equation gives

$$3\left(\frac{11}{4}\right) + 2(-1) + \frac{3}{4} = \frac{33}{4} - 2 + \frac{3}{4} = \frac{33}{4} - \frac{8}{4} + \frac{3}{4} = \frac{28}{4} = 7$$

Again, the first equation is satisfied. Trying the second equation gives

$$4\left(\frac{11}{4}\right) + 9(-1) = 11 - 9 = 2$$

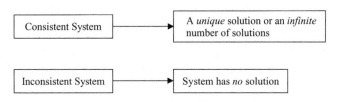

Fig. 1-1. Description of solution possibilities.

This time the second equation is also satisfied. Finally, the third equation works out to be

$$\frac{11}{4} + 5\,(-1) - 3\left(-\frac{3}{4}\right) = \frac{11}{4} - 5 + \frac{9}{4} = \left(\frac{11}{4} + \frac{9}{4}\right) - 5 = \frac{20}{4} - 5 = 5 - 5 = 0$$

This shows that the third equation is satisfied as well. Therefore we conclude that

$$(x, y, z) = \left(\frac{11}{4}, -1, -\frac{3}{4}\right)$$

is a solution to the system.

Consistent and Inconsistent Systems

When at least one solution exists for a given system of linear equations, we call that system *consistent*. If no solution exists, the system is called *inconsistent*. The solution to a system is not necessarily unique. A consistent system either has a unique solution or it can have an infinite number of solutions. We summarize these ideas in Fig. 1-1.

If a consistent system has an infinite number of solutions, if we can define a solution in terms of some extra parameter t, we call this a *parametric solution*.

Matrix Representation of a System of Equations

It is convenient to write down the coefficients and scalars in a linear system of equations as a rectangular array of numbers called a *matrix*. Each row in the array corresponds to one equation. For a system with m equations in n unknowns, there will be m rows in the matrix.

The array will have $n + 1$ columns. Each of the first n columns is used to write the coefficients that multiply each of the unknown variables. The last column is used to write the numbers found on the right-hand side of the equations. Consider the set of equations used in the last example:

$$3x + 2y - z = 7$$
$$4x + 9y = 2$$
$$x + 5y - 3z = 0$$

The matrix used to represent this system is

$$\left[\begin{array}{ccc|c} 3 & 2 & -1 & 7 \\ 4 & 9 & 0 & 2 \\ 1 & 5 & -3 & 0 \end{array}\right]$$

We represent this set of equations

$$2x + y = -7$$
$$x - 5y = 12$$

by the matrix

$$\left[\begin{array}{cc|c} 2 & 1 & -7 \\ 1 & -5 & 12 \end{array}\right]$$

One way we can characterize a matrix is by the number of rows and columns it has. A matrix with m rows and n columns is referred to as an $m \times n$ matrix. Sometimes matrices are *square*, meaning that the number of rows equals the number of columns.

We refer to a given element found in a matrix by identifying its row and column position. This can be done using the notation (i, j) to refer to the element located at row i and column j. Rows are numbered starting with 1 at the top of the matrix, increasing as we move down the matrix. Columns are numbered starting with 1 on the left-hand side.

An alternative method of identifying elements in a matrix is to use a subscript notation. Matrices are often identified with italicized or bold capital letters. So A, B, C or $\mathbf{A}$, $\mathbf{B}$, $\mathbf{C}$ can be used as labels to identify matrices. The corresponding small letter is then used to identify individual elements of the matrix, with subscripts indicating the row and column where the term is located. For a matrix A, we can use a_{ij} to identify the element located at the row and column position (i, j).

As an example, consider the 3×4 matrix

$$B = \begin{bmatrix} -1 & 2 & 7 & 5 \\ 0 & 2 & -1 & 0 \\ 8 & 17 & 21 & -6 \end{bmatrix}$$

The element located at row 2 and column 3 of this matrix can be indicated by writing $(2, 3)$ or b_{23}. This number is

$$b_{23} = -1$$

The element located at row 3 and column 2 is

$$b_{32} = 17$$

The subscript notation is shown in Fig. 1-2.

A matrix that includes the entire linear system is called an *augmented matrix*. We can also make a matrix that is made up only of the coefficients that multiply the unknown variables. This is known as the *coefficient matrix*. For the system

$$5x - y + 9z = 2$$
$$4x + 2y - z = 18$$
$$x + y + 3z = 6$$

the coefficient matrix is

Fig. 1-2. The indexing of an element found at row i and column j of a matrix.

$$A = \begin{bmatrix} 5 & -1 & 9 \\ 4 & 2 & -1 \\ 1 & 1 & 3 \end{bmatrix}$$

We can find a solution to a linear system of equations by applying a set of *elementary operations* to the augmented matrix.

Solving a System Using Elementary Operations

There exist three *elementary operations* that can be applied to a system of linear equations without fundamentally changing that system. These are

- Exchange two rows of the matrix.
- Replace a row by a scalar multiple of itself, as long as the scalar is nonzero.
- Replace one row by adding the scalar multiple of another row.

Let's introduce some shorthand notation to describe these operations and demonstrate using the matrix

$$M = \begin{bmatrix} 2 & -1 & 5 \\ 1 & 33 & 6 \\ 17 & 4 & 8 \end{bmatrix}$$

To indicate the exchange of rows 2 and 3, we write

$$R_2 \leftrightarrow R_3$$

This transforms the matrix as follows:

$$\begin{bmatrix} 2 & -1 & 5 \\ 1 & 33 & 6 \\ 17 & 4 & 8 \end{bmatrix} \rightarrow \begin{bmatrix} 2 & -1 & 5 \\ 17 & 4 & 8 \\ 1 & 33 & 6 \end{bmatrix}$$

Now let's consider the operation where we replace a row by a scalar multiple of itself. Let's say we wanted to replace the first row in the following way:

$$2R_1 \rightarrow R_1$$

The matrix would be transformed as

$$\begin{bmatrix} 2 & -1 & 5 \\ 1 & 33 & 6 \\ 17 & 4 & 8 \end{bmatrix} \rightarrow \begin{bmatrix} 4 & -2 & 10 \\ 1 & 33 & 6 \\ 17 & 4 & 8 \end{bmatrix}$$

In the third type of operation, we replace a selected row by adding a scalar multiple of a different row. Consider

$$-2R_2 + R_1 \rightarrow R_1$$

The matrix becomes

$$\begin{bmatrix} 2 & -1 & 5 \\ 1 & 33 & 6 \\ 17 & 4 & 8 \end{bmatrix} \rightarrow \begin{bmatrix} 0 & -67 & -7 \\ 1 & 33 & 6 \\ 17 & 4 & 8 \end{bmatrix}$$

The solution to the system is obtained when this set of operations brings the matrix into *triangular form*. This type of elimination is sometimes known as *Gaussian elimination*.

Triangular Matrices

Generally, the goal of performing the elementary operations on a system is to get it in a triangular form. A system that is in an upper triangular form is

$$B = \begin{bmatrix} 5 & -1 & 1 & | & 11 \\ 0 & 2 & -1 & | & 2 \\ 0 & 0 & 3 & | & 12 \end{bmatrix}$$

This augmented matrix represents the equations

$$5x - y + z = 11$$
$$2y - z = 2$$
$$3z = 12$$

A solution for the last variable can be found by inspection. In this example, we see that $z = 4$.

To find the values of the other variables, we use *back substitution*. We substitute the value we have found into the equation immediately above it. In this case, insert the value found for z into the second equation. This allows us to solve for y:

$$2y - z = 2, \quad z = 4$$
$$\Rightarrow 2y - 4 = 2$$
$$\therefore y = 3$$

(Note that the symbol $\therefore$ is shorthand for *therefore*.) Each time you apply back substitution, you obtain an equation that has only one unknown variable. Now we can substitute the values $y = 3$ and $z = 4$ into the first equation to solve for the final unknown, which is x:

$$5x - 3 + 4 = 11$$
$$\Rightarrow 5x = 10$$
$$\therefore x = 2$$

A system that is triangular is said to be in *echelon* form. Let's illustrate the complete solution of a system of linear equations using the elementary row operations (see Fig. 1-3).

PIVOTS

Once a system has been reduced, we call the coefficient of the first unknown in each row a *pivot*. For example, in the reduced system

$$3x - 5y + z = 7$$
$$8y - z = 12$$
$$-18z = 11$$

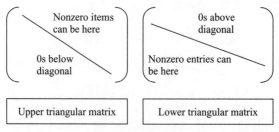

Fig. 1-3. An illustration of an upper triangular matrix, which has 0s below the diagonal, and a lower triangular matrix, which has 0s above the diagonal.

the pivots are 3 for the first row, 8 for the second row, and -18 for the last row. This is also true when representing the system with a matrix. For instance, if the matrix

$$A = \begin{bmatrix} -2 & 11 & 0 & 19 \\ 0 & 16 & -1 & 7 \\ 0 & 0 & 11 & 21 \\ 0 & 0 & 0 & 14 \end{bmatrix}$$

is a coefficient matrix for some system of linear equations, then the pivots are $-2, 16, 11$, and 14.

MORE ON ROW ECHELON FORM

An echelon system has two characteristics:

- Any rows that contain all zeros are found at the bottom of the matrix.
- The first nonzero entry on each row is found to the right of the first nonzero entry in the preceding row.

An echelon system generally has the form

$$a_{11}x_1 + a_{12}x_2 + a_{13}x_3 + \cdots + a_{1n}x_n = b_1$$

$$a_{2j_2}x_{j_2} + a_{2j_2+1}x_{j_2+1} + \cdots + a_{2n}x_n = b_2$$

$$\vdots$$

$$a_{2j_r}x_{jr} + \cdots + a_{rn}x_n = b_r$$

The pivot variables are $x_1, x_{j_2}, \ldots, x_{j_r}$ and the coefficients multiplying each pivot variable are not zero. We also have $r \leq n$.

EXAMPLE 1-2
The following matrices are in echelon form:

$$A = \begin{bmatrix} -2 & 1 & 5 \\ 0 & 1 & 9 \\ 0 & 0 & 8 \end{bmatrix}, \qquad B = \begin{bmatrix} 2 & 0 & 1 \\ 0 & 0 & 1 \\ 0 & 0 & 0 \end{bmatrix}, \qquad C = \begin{bmatrix} 0 & 6 & 0 & 1 \\ 0 & 0 & -2 & 1 \\ 0 & 0 & 0 & 5 \end{bmatrix}$$

The pivots in matrix A are $-2, 1$, and 8. In matrix B, the pivots are 2 and 1, while in matrix C the pivots are 6, -2, and 5.

If a system is brought into row echelon form and it has n equations with n unknowns (so it will be written in a triangular form), then it has a unique solution. If there are m unknowns and n equations with $m > n$ then the values of $m - n$ of the variables are arbitrary. This means that there are an infinite number of solutions.

CANONICAL FORM

If the pivot in each row is a 1 and the pivot is the only nonzero entry in its column, we say that the matrix or system is in a *row canonical form*.
The matrix

$$A = \begin{bmatrix} 1 & 0 & 0 & 0 \\ 0 & 1 & 0 & -6 \\ 0 & 0 & 1 & 2 \\ 0 & 0 & 0 & 0 \end{bmatrix}$$

is in a row canonical form because all of the pivots are equal to 1 and they are the only nonzero elements in their respective columns. The matrix

$$B = \begin{bmatrix} 1 & 0 & 8 \\ 0 & 0 & -2 \\ 0 & 0 & 0 \end{bmatrix}$$

is not in a row canonical form because there is a nonzero entry above the pivot in the second row.

ROW EQUIVALENCE

If a matrix B can be obtained from a matrix A by using a series of elementary row operations, then we say the matrices are row equivalent. This is indicated using the notation

$$A \sim B$$

RANK OF A MATRIX

The rank of a matrix is the number of pivots in the echelon form of the matrix.

EXAMPLE 1-3

The rank of

$$A = \begin{bmatrix} 1 & 0 & 0 & 0 \\ 0 & 1 & 0 & -6 \\ 0 & 0 & 1 & 2 \\ 0 & 0 & 0 & 0 \end{bmatrix}$$

is 3, because the matrix is in echelon form and has three pivots. The rank of

$$B = \begin{bmatrix} -2 & 0 & 7 \\ 0 & 4 & 5 \\ 0 & 0 & 1 \end{bmatrix}$$

is 3. The matrix is in echelon form, and it has three pivots, $-2, 4, 1$.

EXAMPLE 1-4

Find a solution to the system

$$5x_1 + 2x_2 - 3x_3 = 4$$

$$x_1 - x_2 + 2x_3 = -1$$

SOLUTION 1-4

There are two equations in three unknowns. This means that we can find a solution in terms of a single parametric variable we call t. There are an infinite number of solutions because unless more constraints have been stated for the problem, we can choose any value for t.

We can eliminate x_1 from the second row by using $R_1 - 5R_2 \rightarrow R_2$, which gives

$$5x_1 + 2x_2 - 3x_3 = 4$$

$$7x_2 - 13x_3 = 9$$

From the second equation, we obtain

$$x_2 = \frac{1}{7}(9 + 13x_3)$$

We substitute this expression into the first equation and solve for x_1 in terms of x_3. We find that

$$x_1 = \frac{2}{7} - \frac{1}{7}x_3$$

Now we set

$$x_3 = t$$

where t is a parameter. With no further information, there are an infinite number of solutions because t can be anything. For example, if $t = 5$ then the solution is

$$x_1 = \frac{-3}{7}, \qquad x_2 = \frac{74}{7}, \qquad x_3 = 5$$

But $t = 0$ is also a valid solution, giving

$$x_1 = \frac{2}{7}, \qquad x_2 = \frac{9}{7}, \qquad x_3 = 0$$

We could continue choosing various values of t. Instead we write

$$x_1 = \frac{2}{7} - \frac{1}{7}t, \qquad x_2 = \frac{9}{7} + \frac{13}{7}t, \qquad x_3 = t$$

EXAMPLE 1-5
Find a solution to the system

$$3x - 7y + 2z = 1$$
$$x + y - 5z = 15$$
$$-x + 2y - 3z = 4$$

SOLUTION 1-5
First we write down the augmented matrix. Arranging the coefficients on the left side and the constants on the right, we have

$$A = \left[\begin{array}{ccc|c} 3 & -7 & 2 & 1 \\ 1 & 1 & -5 & 15 \\ -1 & 2 & -3 & 4 \end{array} \right]$$

The first step in solving a linear system is to identify a *pivot*. The idea is to eliminate all terms in the matrix below the pivot so that we can write the matrix in an upper triangular form.

In this case, we take $a_{11} = 3$ as the first pivot and eliminate all coefficients below this value. Notice that we can eliminate the first coefficient in the third row by using the elementary row operation

$$R_1 + 3R_3 \rightarrow R_3$$

This will transform the matrix in the following way:

$$\begin{bmatrix} 3 & -7 & 2 & 1 \\ 1 & 1 & -5 & 15 \\ -1 & 2 & -3 & 4 \end{bmatrix} \xrightarrow{R_1 + 3R_3 \rightarrow R_3} \begin{bmatrix} 3 & -7 & 2 & 1 \\ 1 & 1 & -5 & 15 \\ 0 & -1 & -7 & 13 \end{bmatrix}$$

Next, we eliminate the remaining value below the first pivot, which is the first element in the second row. We can do this with

$$R_1 - 3R_2 \rightarrow R_2$$

This gives

$$\begin{bmatrix} 3 & -7 & 2 & 1 \\ 1 & 1 & -5 & 15 \\ 0 & -1 & -7 & 13 \end{bmatrix} \xrightarrow{R_1 - 3R_2 \rightarrow R_2} \begin{bmatrix} 3 & -7 & 2 & 1 \\ 0 & -10 & 17 & -44 \\ 0 & -1 & -7 & 19 \end{bmatrix}$$

At this point we have done all we can with the first pivot. To identify the next pivot, we move down one row and then move right one column. In this case, the next pivot in the matrix

$$\begin{bmatrix} 3 & -7 & 2 & 1 \\ 0 & -10 & 17 & -44 \\ 0 & -1 & -7 & 19 \end{bmatrix}$$

is

$$a_{22} = -10$$

We use the second pivot to eliminate the coefficient found immediately below it with the elementary row operation

$$R_2 - 10R_3 \rightarrow R_3$$

This allows us to rewrite the matrix in the following way:

$$\left[\begin{array}{ccc|c} 3 & -7 & 2 & 1 \\ 0 & -10 & 17 & -44 \\ 0 & -1 & -7 & 19 \end{array}\right] \xrightarrow{R_2-10R_3 \rightarrow R_3} \left[\begin{array}{ccc|c} 3 & -7 & 2 & 1 \\ 0 & -10 & 17 & -44 \\ 0 & 0 & 87 & -174 \end{array}\right]$$

Now the matrix is triangular. Or we can say it is in echelon form. This means that

- Row 1 has three nonzero coefficients.
- Row 2 has two nonzero coefficients: the first nonzero coefficient is to the right of the column where the first nonzero coefficient is located in row 1.
- Row 3 has one nonzero coefficient: it is also to the right of the first nonzero coefficient in row 2.

The pivots are 3, -10, and 87. This allows us to solve for the last variable immediately. The equation is

$$87z = -174$$
$$\Rightarrow z = \frac{-174}{87} = -2$$

With $z = -2$, we can use back substitution to solve for the other variables. We move up one row, and the equation is

$$-10y + 17z = -44$$

Making the substitution $z = -2$ allows us to write this as

$$-10y + 17(-2) = -10y - 34 = -44$$

Now add 34 to both sides, which gives

$$-10y = -10$$

Dividing both sides by -10, we get

$$y = 1$$

The final equation in this system is

$$3x - 7y + 2z = 1$$

Substitution of $y = 1, z = -2$ allows us to write the left-hand side as

$$3x - 7(1) + 2(-2) = 3x - 7 - 4 = 3x - 11$$

Setting this equal to the right-hand side gives

$$3x - 11 = 1$$
$$\Rightarrow 3x = 12$$

Now dividing both sides by 3, we find that

$$x = 4$$

The complete solution is given by

$$(x, y, z) = (4, 1, -2)$$

EXAMPLE 1-6
Find a solution to the system

$$x - 3y + z = 2$$
$$5x + 2y - 4z = 8$$
$$-x + 3y + z = -1$$

SOLUTION 1-6
The augmented matrix for this system is

$$\left[\begin{array}{ccc|c} 1 & -3 & 1 & 2 \\ 5 & 2 & -4 & 8 \\ -1 & 3 & 1 & -1 \end{array}\right]$$

We select the term located at $(1, 1)$ as the first pivot. We proceed to eliminate all terms below the pivot, using elementary row operations. To begin, add the first row to the third.

$$R_1 + R_3 \rightarrow R_3$$

This gives

$$\begin{bmatrix} 1 & -3 & 1 & | & 2 \\ 5 & 2 & -4 & | & 8 \\ 0 & 0 & 2 & | & 1 \end{bmatrix}$$

Next we wish to eliminate the term located at position $(2, 1)$. We can do this with the operation

$$-5R_1 + R_2 \rightarrow R_2$$

The augmented matrix becomes

$$\begin{bmatrix} 1 & -3 & 1 & | & 2 \\ 0 & 17 & -9 & | & -2 \\ 0 & 0 & 2 & | & 1 \end{bmatrix}$$

The matrix is now in an upper triangular form. For the last variable the equation that described the bottom row is

$$2z = 1$$

and so we have

$$z = \frac{1}{2}$$

Back substitution into the next row gives

$$17y = 9z - 2$$

$$\Rightarrow y = \frac{5}{34}$$

Now we use back substitution of the values found for y and z into the equation described by the top row to solve for x. The equation is

$$x = 3y - z + 2 = 3\left(\frac{5}{34}\right) - \frac{1}{2} + 2 = \frac{15}{34} - \frac{17}{34} + \frac{68}{34} = \frac{66}{34} = \frac{33}{17}$$

While ideally we want to get the matrix in triangular form, this is not always necessary. We show this in the next example.

EXAMPLE 1-7
Use Gaussian elimination to find a solution to the following system:

$$2y - z = 1$$
$$-x + 2y - z = 0$$
$$x - 4y + z = 2$$

SOLUTION 1-7
The augmented matrix is

$$\left[\begin{array}{ccc|c} 0 & 2 & -1 & 1 \\ -1 & 2 & -1 & 0 \\ 1 & -4 & 1 & 2 \end{array}\right]$$

The first pivot position contains a zero. We exchange rows 1 and 3 to move a nonzero value into the first pivot.

$$R_1 \leftrightarrow R_3$$

This gives

$$\left[\begin{array}{ccc|c} 1 & -4 & 1 & 2 \\ -1 & 2 & -1 & 0 \\ 0 & 2 & -1 & 1 \end{array}\right]$$

There is only one term to eliminate below the first pivot. We add the first row to the second row:

$$R_1 + R_2 \rightarrow R_2$$

and the result is

$$\left[\begin{array}{ccc|c} 1 & -4 & 1 & 2 \\ 0 & -2 & 0 & 2 \\ 0 & 2 & -1 & 1 \end{array}\right]$$

The second row tells us that

$$-2y = 2$$

Therefore, $y = -1$. We substitute this value into the third equation:

$$z = 2y - 1 = -2 - 1 = -3$$

We can then find x, using the equation in the top row:

$$x = 4y - z + 2 = -4 + 3 + 2 = 1$$

Elementary Matrices

When dealing with a square $n \times n$ system, elementary row operations can be represented by a set of matrices called *elementary matrices*. In this example we focus on the 3×3 case. To create an elementary matrix, write down a 3×3 matrix that has 1s on the diagonal and 0s everywhere else:

$$I_3 = \left[\begin{array}{ccc} 1 & 0 & 0 \\ 0 & 1 & 0 \\ 0 & 0 & 1 \end{array}\right]$$

We'll see in a minute how to use these matrices to implement row operations for a given matrix. Right now let's concentrate on representing each type of operation.

REPRESENTATION OF A ROW EXCHANGE USING ELEMENTARY MATRICES

To represent the operation $R_m \leftrightarrow R_n$, we simply exchange the corresponding rows in the matrix I_n. For example, in the 3×3 case, to exchange rows 1 and

2, we write

$$\begin{bmatrix} 0 & 1 & 0 \\ 1 & 0 & 0 \\ 0 & 0 & 1 \end{bmatrix}$$

To exchange rows 1 and 3, we write

$$\begin{bmatrix} 0 & 0 & 1 \\ 0 & 1 & 0 \\ 1 & 0 & 0 \end{bmatrix}$$

and to exchange rows 2 and 3, we have

$$\begin{bmatrix} 1 & 0 & 0 \\ 0 & 0 & 1 \\ 0 & 1 & 0 \end{bmatrix}$$

REPLACING A ROW BY A MULTIPLE OF ITSELF

To implement the operation $\alpha R_m \to R_m$, where R_m is row m and α is a scalar in the 3×3 case, we use the following elementary matrices.

We represent the multiplication of the first row of a matrix by α with

$$\begin{bmatrix} \alpha & 0 & 0 \\ 0 & 1 & 0 \\ 0 & 0 & 1 \end{bmatrix}$$

For the second row we use

$$\begin{bmatrix} 1 & 0 & 0 \\ 0 & \alpha & 0 \\ 0 & 0 & 1 \end{bmatrix}$$

and for the third row we have

$$\begin{bmatrix} 1 & 0 & 0 \\ 0 & 1 & 0 \\ 0 & 0 & \alpha \end{bmatrix}$$

REPLACE ONE ROW BY ADDING THE SCALAR MULTIPLE OF ANOTHER ROW

The last type of operation is slightly more complicated. Suppose that we want to write down the elementary matrix that corresponds to the operation

$$\alpha R_i + \beta R_j \rightarrow R_j$$

where R_i is row i and R_j is row j. To do this, we start with I_n and modify row j in the following way:

- Replace the element in column i by α.
- Replace the element in column j by β.

EXAMPLE 1-8

For a 3×3 matrix A, write down the three elementary matrices that correspond to the row operations

- $R_2 \leftrightarrow R_3$
- $4R_2 \rightarrow R_2$
- $3R_1 + R_3 \rightarrow R_3$

SOLUTION 1-8

We start with I_3

$$I_3 = \begin{bmatrix} 1 & 0 & 0 \\ 0 & 1 & 0 \\ 0 & 0 & 1 \end{bmatrix}$$

The row operation $R_2 \leftrightarrow R_3$ is represented by swapping rows 2 and 3 in the I_3 matrix

$$\begin{bmatrix} 1 & 0 & 0 \\ 0 & 0 & 1 \\ 0 & 1 & 0 \end{bmatrix}$$

To represent $4R_2 \rightarrow R_2$, we replace the second row of I_3 with

$$\begin{bmatrix} 1 & 0 & 0 \\ 0 & 4 & 0 \\ 0 & 0 & 1 \end{bmatrix}$$

Now we consider $3R_1 + R_3 \rightarrow R_3$. We will modify row 3, which is the destination row, in the I_3 matrix. We will need to replace the element in the first column, which is a 0, with a 3. The element in the third column is unchanged because the scalar multiple is 1, and so we use

$$\begin{bmatrix} 1 & 0 & 0 \\ 0 & 1 & 0 \\ 3 & 0 & 1 \end{bmatrix}$$

EXAMPLE 1-9
Represent the operations

$$2R_1 - R_2 \rightarrow R_2 \quad \text{and} \quad 4R_2 + 6R_3 \rightarrow R_3$$

with elementary matrices in a 3×3 system.

SOLUTION 1-9
To represent $2R_1 - R_2 \rightarrow R_2$, we will modify row 2 of I_3. We replace the element in the first column with a 2, and change the element in the second column with a -1. This gives

$$\begin{bmatrix} 1 & 0 & 0 \\ 2 & -1 & 0 \\ 0 & 0 & 1 \end{bmatrix}$$

To represent the second operation, we replace the third row of I_3. The operation is $4R_2 + 6R_3 \rightarrow R_3$, and so we replace the element at the second column with a 4, and the element in the third column with a 6, which results in the matrix

$$\begin{bmatrix} 1 & 0 & 0 \\ 0 & 1 & 0 \\ 0 & 4 & 6 \end{bmatrix}$$

EXAMPLE 1-10
For a 4×4 matrix, find the elementary matrix that represents

$$-2R_2 + 5R_4 \rightarrow R_4$$

SOLUTION 1-10

To construct an elementary matrix, we begin with a matrix with 1s along the diagonal and 0s everywhere else. For a 4×4 matrix, we use

$$I_4 = \begin{bmatrix} 1 & 0 & 0 & 0 \\ 0 & 1 & 0 & 0 \\ 0 & 0 & 1 & 0 \\ 0 & 0 & 0 & 1 \end{bmatrix}$$

The destination row is the fourth row, and so we will modify the fourth row of I_4. The operation involves adding -2 times the second row to 5 times the fourth row. And so we will replace the element located in the second column by -2 and the element in the fourth column by 5, which gives

$$\begin{bmatrix} 1 & 0 & 0 & 0 \\ 0 & 1 & 0 & 0 \\ 0 & 0 & 1 & 0 \\ 0 & -2 & 0 & 5 \end{bmatrix}$$

Implementing Row Operations with Elementary Matrices

Row operations are implemented with elementary matrices using *matrix multiplication*. We will explore matrix multiplication in detail in the next chapter, but it turns out that matrix multiplication using an elementary matrix is particularly simple. For now, we will show how to do this for 2×2 and 3×3 matrices.

MATRIX MULTIPLICATION BY A 2×2 ELEMENTARY MATRIX

Let E be an elementary matrix and A be an arbitrary 2×2 matrix given by

$$A = \begin{bmatrix} a & b \\ c & d \end{bmatrix}$$

We have two cases to consider, operations on the first and second rows. An arbitrary operation on the first row is represented by

$$E_1 = \begin{bmatrix} \alpha & \beta \\ 0 & 1 \end{bmatrix}$$

The product $E_1 A$ is given by

$$E_1 A = \begin{bmatrix} \alpha & \beta \\ 0 & 1 \end{bmatrix} \begin{bmatrix} a & b \\ c & d \end{bmatrix} = \begin{bmatrix} \alpha a + \beta c & \alpha b + \beta d \\ c & d \end{bmatrix}$$

An operation on row 2 is given by

$$E_2 = \begin{bmatrix} 1 & 0 \\ \alpha & \beta \end{bmatrix}$$

and the product $E_2 A$ is

$$E_2 A = \begin{bmatrix} 1 & 0 \\ \alpha & \beta \end{bmatrix} \begin{bmatrix} a & b \\ c & d \end{bmatrix} = \begin{bmatrix} a & b \\ \alpha a + \beta c & \alpha b + \beta d \end{bmatrix}$$

EXAMPLE 1-11
Consider the matrix

$$A = \begin{bmatrix} -2 & 5 \\ 4 & 11 \end{bmatrix}$$

Implement the row operations $2R_1 \rightarrow R_1$ and $-3R_1 + R_2 \rightarrow R_2$ using elementary matrices.

SOLUTION 1-11
The operation $2R_1 \rightarrow R_1$ is represented by the elementary matrix

$$E = \begin{bmatrix} 2 & 0 \\ 0 & 1 \end{bmatrix}$$

Using the formulas developed above, we have

$$EA = \begin{bmatrix} 2 & 0 \\ 0 & 1 \end{bmatrix} \begin{bmatrix} -2 & 5 \\ 4 & 11 \end{bmatrix} = \begin{bmatrix} (2)(-2)+(0)(4) & (2)(5)+(0)(11) \\ 4 & 11 \end{bmatrix}$$

$$= \begin{bmatrix} -4+0 & 10+0 \\ 4 & 11 \end{bmatrix} = \begin{bmatrix} -4 & 10 \\ 4 & 11 \end{bmatrix}$$

The elementary matrix that represents $-3R_1 + R_2 \rightarrow R_2$ is found to be

$$E = \begin{bmatrix} 1 & 0 \\ -3 & 1 \end{bmatrix}$$

The product is

$$EA = \begin{bmatrix} 1 & 0 \\ -3 & 1 \end{bmatrix} \begin{bmatrix} -2 & 5 \\ 4 & 11 \end{bmatrix} = \begin{bmatrix} -2 & 5 \\ (-3)(-2)+(1)(4) & (-3)(5)+(1)(11) \end{bmatrix}$$

$$= \begin{bmatrix} -2 & 5 \\ 6+4 & -15+11 \end{bmatrix} = \begin{bmatrix} -2 & 5 \\ 10 & -4 \end{bmatrix}$$

ROW OPERATIONS ON A 3 × 3 MATRIX

Row operations on 3×3 matrix A are best shown with example. The multiplication techniques are similar to those used above.

EXAMPLE 1-12
Consider the matrix

$$A = \begin{bmatrix} 7 & -2 & 3 \\ 0 & 1 & 4 \\ -2 & 3 & 5 \end{bmatrix}$$

Implement the row operations $2R_2 \rightarrow R_2$, $R_1 \leftrightarrow R_3$, $-4R_1 + R_2 \rightarrow R_2$ using elementary matrices.

SOLUTION 1-12
The elementary matrix that corresponds to $2R_2 \rightarrow R_2$ is given by

$$E_1 = \begin{bmatrix} 1 & 0 & 0 \\ 0 & 2 & 0 \\ 0 & 0 & 1 \end{bmatrix}$$

The operation is implemented by computing the product of this matrix with A:

$$E_1 A = \begin{bmatrix} 1 & 0 & 0 \\ 0 & 2 & 0 \\ 0 & 0 & 1 \end{bmatrix} \begin{bmatrix} 7 & -2 & 3 \\ 0 & 1 & 4 \\ -2 & 3 & 5 \end{bmatrix}$$

$$= \begin{bmatrix} 7 & -2 & 3 \\ (0)(7)+(2)(0)+(0)(-2) & (0)(7)+(2)(1)+(0)(-2) & (0)(7)+(2)(4)+(0)(-2) \\ -2 & 3 & 5 \end{bmatrix}$$

$$= \begin{bmatrix} 7 & -2 & 3 \\ 0 & 2 & 8 \\ -2 & 3 & 5 \end{bmatrix}$$

The swap operation $R_1 \leftrightarrow R_3$ can be implemented with the matrix

$$E_2 = \begin{bmatrix} 0 & 0 & 1 \\ 0 & 1 & 0 \\ 1 & 0 & 0 \end{bmatrix}$$

In this case rows 1 and 3 have been changed. So we will multiply both rows in this case. The result is

$$E_2 A = \begin{bmatrix} 0 & 0 & 1 \\ 0 & 1 & 0 \\ 1 & 0 & 0 \end{bmatrix} \begin{bmatrix} 7 & -2 & 3 \\ 0 & 1 & 4 \\ -2 & 3 & 5 \end{bmatrix}$$

$$= \begin{bmatrix} (0)(7)+(0)(0)+(1)(-2) & (0)(-2)+(0)(1)+(1)(3) & (0)(3)+(0)(4)+(1)(5) \\ 0 & 1 & 4 \\ (1)(7)+(0)(0)+(0)(-2) & (1)(-2)+(0)(1)+(0)(3) & (1)(3)+(0)(4)+(0)(5) \end{bmatrix}$$

$$= \begin{bmatrix} 0+0-2 & 0+0+3 & 0+0+5 \\ 0 & 1 & 4 \\ 7+0+0 & -2+0+0 & 3+0+0 \end{bmatrix}$$

$$= \begin{bmatrix} -2 & 3 & 5 \\ 0 & 1 & 4 \\ 7 & -2 & 3 \end{bmatrix}$$

Finally, we can implement the operation $-4R_1 + R_2 \rightarrow R_2$, using the elementary matrix

$$E_3 = \begin{bmatrix} 1 & 0 & 0 \\ -4 & 1 & 0 \\ 0 & 0 & 1 \end{bmatrix}$$

We find

$$E_3 A = \begin{bmatrix} 1 & 0 & 0 \\ -4 & 1 & 0 \\ 0 & 0 & 1 \end{bmatrix} \begin{bmatrix} 7 & -2 & 3 \\ 0 & 1 & 4 \\ -2 & 3 & 5 \end{bmatrix}$$

$$= \begin{bmatrix} 7 & -2 & 3 \\ (-4)(7)+(1)(0) & (-4)(-2)+(1)(1) & (-4)(3)+(1)(4) \\ -2 & 3 & 5 \end{bmatrix}$$

$$= \begin{bmatrix} 7 & -2 & 3 \\ -28+0 & 8+1 & -12+4 \\ -2 & 3 & 5 \end{bmatrix} = \begin{bmatrix} 7 & -2 & 3 \\ -28 & 9 & -8 \\ -2 & 3 & 5 \end{bmatrix}$$

Homogeneous Systems

A *homogeneous system* is a linear system with all zeros on the right-hand side. In general, it is a system of the form

$$a_{11}x_1 + a_{12}x_2 + \cdots + a_{1n}x_n = 0$$
$$a_{21}x_1 + a_{22}x_2 + \cdots + a_{2n}x_n = 0$$
$$\vdots$$
$$a_{m1}x_1 + a_{m2}x_2 + \cdots + a_{mn}x_n = 0$$

When a system is put in echelon form, if the system has more unknowns than equations, then it has a nonzero solution. A system in echelon form with n equations and n unknowns has only the zero solution, meaning that only $(x, y, z) = (0, 0, 0)$ solves the system.

EXAMPLE 1-13
Determine if the system

$$2x - 8y + z = 0$$
$$x + y - z = 0$$
$$3x + 3y + 2z = 0$$

has a nonzero solution.

SOLUTION 1-13
We bring the system of equations into echelon form. First we perform the row operation $-3R_1 + 2R_3 \rightarrow R_3$, which results in

$$2x - 8y + z = 0$$
$$x + y - z = 0$$
$$30! \quad -\underbrace{(27y)} - z = 0$$

Next we apply $R_1 - 2R_2 \rightarrow R_2$, which gives

$$2x - 8y + z = 0$$
$$-10y + 3z = 0$$
$$27y - z = 0$$

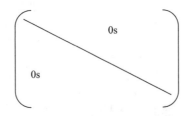

The system is now in row echelon form. There are three equations and three unknowns, and therefore the system has only the zero solution.

Gauss-Jordan Elimination

In Gauss-Jordan elimination, there are 0s both above and below each pivot. This way of reducing a matrix is less efficient (see Fig. 1-4).

Os

Os

Fig. 1-4. In Gauss-Jordan elimination, we put 0s above and below the pivots.

EXAMPLE 1-14
Reduce the matrix

$$A = \begin{bmatrix} 1 & 3 & 0 & -1 \\ 2 & 5 & 3 & -2 \\ 3 & 7 & 5 & -4 \end{bmatrix}$$

to row canonical form using Gauss-Jordan elimination.

SOLUTION 1-14
We choose the entry in row 1 and column 1 as the first pivot. First we eliminate all entries below this number. To eliminate the entry in the second row, we use $-2R_1 + R_2 \rightarrow R_2$
This gives

$$\begin{bmatrix} 1 & 3 & 0 & -1 \\ 0 & -1 & 3 & 0 \\ 3 & 7 & 5 & -4 \end{bmatrix}$$

Next we use $-3R_1 + R_3 \rightarrow R_3$ to eliminate the next entry below the first pivot, and we obtain

$$\begin{bmatrix} 1 & 3 & 0 & -1 \\ 0 & -1 & 3 & 0 \\ 0 & -2 & 5 & -1 \end{bmatrix}$$

Now we select the second entry in row 2 as the next pivot. We eliminate the entry directly below this value using $-2R_2 + R_3 \rightarrow R_3$ and we have

$$\begin{bmatrix} 1 & 3 & 0 & -1 \\ 0 & -1 & 3 & 0 \\ 0 & 0 & -1 & -1 \end{bmatrix}$$

In Gauss-Jordan elimination, we want to eliminate all entries above the pivot as well. So we use $3R_2 + R_1 \rightarrow R_1$, which gives

$$\begin{bmatrix} 1 & 0 & 9 & -1 \\ 0 & -1 & 3 & 0 \\ 0 & 0 & -1 & -1 \end{bmatrix}$$

Next we eliminate terms above the a_{33} entry. First we eliminate the term immediately above by carrying out $3R_3 + R_2 \rightarrow R_2$, which gives

$$\begin{bmatrix} 1 & 0 & 9 & -1 \\ 0 & -1 & 0 & -3 \\ 0 & 0 & -1 & -1 \end{bmatrix}$$

Now we eliminate the term in the first row above the a_{33} entry using $9R_3 + R_1 \rightarrow$
R_1 and we obtain

$$\begin{bmatrix} 1 & 0 & 0 & -10 \\ 0 & -1 & 0 & -3 \\ 0 & 0 & -1 & -1 \end{bmatrix}$$

To put the matrix in row canonical form, the leftmost entries in each row must
be equal to 1. We divide rows 2 and 3 by -1 to obtain the row canonical form

$$\begin{bmatrix} 1 & 0 & 0 & -10 \\ 0 & 1 & 0 & 3 \\ 0 & 0 & 1 & 1 \end{bmatrix}$$

EXAMPLE 1-15
Solve the system

$$x - 2y + 3z = 1$$
$$x + y + 4z = -1$$
$$2x + 5y + 4z = -3$$

using Gauss-Jordan elimination.

SOLUTION 1-15
The augmented matrix is

$$A = \begin{bmatrix} 1 & -2 & 3 & 1 \\ 1 & 1 & 4 & -1 \\ 2 & 5 & 4 & -3 \end{bmatrix}$$

First we eliminate all terms below the first entry in the first column. $-R_1 + R_2 \rightarrow R_2$ gives

$$\begin{bmatrix} 1 & -2 & 3 & 1 \\ 0 & 3 & 1 & -2 \\ 2 & 5 & 4 & -3 \end{bmatrix}$$

$-2R_1 + R_3 \rightarrow R_3$ changes this to

$$\begin{bmatrix} 1 & -2 & 3 & 1 \\ 0 & 3 & 1 & -2 \\ 0 & 9 & -2 & -5 \end{bmatrix}$$

Now we eliminate terms above and below the second entry in the second column. First we use $-3R_2 + R_3 \rightarrow R_3$ and find

$$\begin{bmatrix} 1 & -2 & 3 & 1 \\ 0 & 3 & 1 & -2 \\ 0 & 0 & -5 & 1 \end{bmatrix}$$

It will be easier to proceed by altering the matrix so that 1s appear in positions a_{22} and a_{33}. We divide row 2 by 3 to obtain

$$\begin{bmatrix} 1 & -2 & 3 & 1 \\ 0 & 1 & \frac{1}{3} & -\frac{2}{3} \\ 0 & 0 & -5 & 1 \end{bmatrix}$$

Now divide row 3 by -5 to obtain

$$\begin{bmatrix} 1 & -2 & 3 & 1 \\ 0 & 1 & \frac{1}{3} & -\frac{2}{3} \\ 0 & 0 & 1 & -\frac{1}{5} \end{bmatrix}$$

Now use $2R_2 + R_1 \rightarrow R_1$ to eliminate the term above a_{22}:

$$\begin{bmatrix} 1 & 0 & \frac{8}{3} & -\frac{1}{3} \\ 0 & 1 & \frac{1}{3} & -\frac{2}{3} \\ 0 & 0 & 1 & -\frac{1}{5} \end{bmatrix}$$

Now we eliminate all terms above a_{33}. We use $-\frac{1}{3}R_3 + R_2 \rightarrow R_2$ to obtain

$$\begin{bmatrix} 1 & 0 & \frac{11}{3} & -\frac{1}{3} \\ 0 & 1 & 0 & -\frac{3}{5} \\ 0 & 0 & 1 & -\frac{1}{5} \end{bmatrix}$$

Now use $-\frac{11}{3}R_3 + R_1 \rightarrow R_1$ and we get the row canonical form we seek

$$\begin{bmatrix} 1 & 0 & 0 & \frac{2}{5} \\ 0 & 1 & 0 & -\frac{3}{5} \\ 0 & 0 & 1 & -\frac{1}{5} \end{bmatrix}$$

From this matrix we immediately read off the solution

$$x = \frac{2}{5}, \qquad y = -\frac{3}{5}, \qquad z = -\frac{1}{5}$$

Quiz

1. Is $(x, y, z) = (8, -13, -6)$ a solution of the system yes

$$4x + 2y + z = 0$$
$$x + y - z = 1$$
$$x + z = 2$$

2. Find a solution to the system

$$-x + y + z = -1$$
$$x + y + z = 1$$
$$x + 2y + z = 2$$

    $x = 1$
    $y = 1$
    $z = 0$

3. Determine whether or not the following system has a solution:

$$x + 2y + z = -1 \qquad No$$
$$3x + 6y - z = 2$$
$$x + z = -2$$

4. Determine whether or not the following system has a solution:

$$-2x + 5y + z = -1$$
$$3x + 6y - z = 2$$
$$y + 8z = -6$$

5. Represent the system

$$5x + 4y + z = -19$$
$$3x + 6y - 2z = 8$$
$$x + 3z = 11$$

with an augmented matrix.

6. For the system

$$3x - 9y + 5z = -11$$
$$3x + 5y - 6z = 18$$
$$5x + z = -2$$

write down the coefficient matrix A.

7. What is the elementary matrix that represents $2R_2 + 7R_3 \rightarrow R_3$ for the matrix

$$A = \begin{bmatrix} -1 & 0 & 4 \\ 5 & 2 & 0 \\ 8 & -7 & 1 \end{bmatrix}$$

$$\begin{bmatrix} 1 & 0 & 0 \\ 0 & 1 & 0 \\ 0 & 2 & 7 \end{bmatrix}$$

8. Find the elementary matrix E that represents $5R_1 + 3R_2 \rightarrow R_2$ for the 2×2 matrix

$$A = \begin{bmatrix} -1 & 3 \\ 4 & 6 \end{bmatrix}$$

and then calculate the product EA.

$$\begin{bmatrix} 1 & 0 \\ 5 & 3 \end{bmatrix}$$

9. Using elementary matrix multiplication, implement $5R_2 \to R_2$ for

$$A = \begin{bmatrix} 2 & 1 & 1 \\ 5 & 6 & -3 \\ 4 & -1 & 1 \end{bmatrix}$$

10. Using elementary matrix multiplication, implement $-2R_2 + R_3 \to R_3$ for

$$A = \begin{bmatrix} 2 & 1 & 1 \\ 5 & 6 & -3 \\ 4 & -1 & 1 \end{bmatrix}$$

11. Use row operations to put the matrix

$$B = \begin{bmatrix} 3 & 2 & -1 & 7 \\ 4 & 0 & 1 & 2 \\ 8 & 7 & -2 & 1 \end{bmatrix}$$

into echelon form and find the rank.

12. Find a parametric solution for the system

$$5w - 2x + y - z = 0$$
$$2w + x + y + z = -1$$
$$-w + 3x - y + 2z = 3$$

13. Use Gauss-Jordan elimination to find the row canonical form of

$$A = \begin{bmatrix} 2 & 2 & -1 & 6 & 4 \\ 4 & 4 & 1 & 10 & 13 \\ 8 & 8 & -1 & 26 & 23 \end{bmatrix}$$

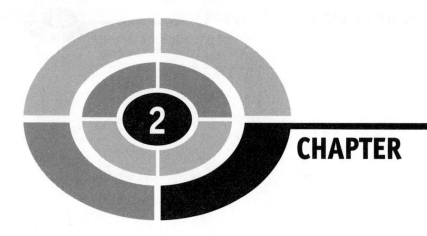

CHAPTER

Matrix Algebra

Basic operations such as addition and multiplication carry over to matrices. However, the operations do not always carry over in a straightforward manner because matrices are more complicated than numbers.

Matrix Addition

If two matrices have the same number of rows and columns then we can add them together to produce a new, third matrix. Suppose that the matrices A and B are $m \times n$ matrices with components a_{ij} and b_{ij}, respectively. We let the matrix C have components c_{ij}. Then we form the sum $C = A + B$ by letting $c_{ij} = a_{ij} + b_{ij}$.

EXAMPLE 2-1
Let

$$A = \begin{bmatrix} 2 & 0 \\ -1 & 4 \end{bmatrix}, \qquad B = \begin{bmatrix} 7 & -1 \\ 2 & 3 \end{bmatrix}$$

Find the matrix $C = A + B$.

SOLUTION 2-1

We find the matrix C by adding the components of the matrices together. We have

$$C = A + B = \begin{bmatrix} 2 & 0 \\ -1 & 4 \end{bmatrix} + \begin{bmatrix} 7 & -1 \\ 2 & 3 \end{bmatrix} = \begin{bmatrix} 2+7 & 0+(-1) \\ -1+2 & 4+3 \end{bmatrix} = \begin{bmatrix} 9 & -1 \\ 1 & 7 \end{bmatrix}$$

Matrix subtraction is done similarly. For example, we could compute

$$C = A - B = \begin{bmatrix} 2 & 0 \\ -1 & 4 \end{bmatrix} - \begin{bmatrix} 7 & -1 \\ 2 & 3 \end{bmatrix} = \begin{bmatrix} 2-7 & 0-(-1) \\ -1-2 & 4-3 \end{bmatrix} = \begin{bmatrix} -5 & 1 \\ -3 & 1 \end{bmatrix}$$

As we shall see in the world of linear algebra, there are two ways to do multiplication. We can multiply a matrix by a number or *scalar* or we can multiply two matrices together.

Scalar Multiplication

Let A be an $m \times n$ matrix with components a_{ij} and let α be a scalar. The scalar multiple of A is formed by multiplying each component a_{ij} by α. Note that α can be real or complex.

EXAMPLE 2-2

Let

$$A = \begin{bmatrix} 4 & -2 & 0 \\ 0 & 1 & 2 \\ 7 & 5 & 9 \end{bmatrix}$$

and suppose that $\alpha = 3$ and $\beta = 2 + 4i$. Find αA and βA.

SOLUTION 2-2

We compute the scalar multiple of A by multiplying each component by the given scalar. For αA we find

$$\alpha A = (3) \begin{bmatrix} 4 & -2 & 0 \\ 0 & 1 & 2 \\ 7 & 5 & 9 \end{bmatrix} = \begin{bmatrix} 3(4) & 3(-2) & 3(0) \\ 3(0) & 3(1) & 3(2) \\ 3(7) & 3(5) & 3(9) \end{bmatrix} = \begin{bmatrix} 12 & -6 & 0 \\ 0 & 3 & 6 \\ 21 & 15 & 27 \end{bmatrix}$$

The calculation of βA proceeds in a similar manner

$$\beta A = (2 + 4i)\begin{bmatrix} 4 & -2 & 0 \\ 0 & 1 & 2 \\ 7 & 5 & 9 \end{bmatrix} = \begin{bmatrix} (2+4i)(4) & (2+4i)(-2) & (2+4i)(0) \\ (2+4i)(0) & (2+4i)(1) & (2+4i)(2) \\ (2+4i)(7) & (2+4i)(5) & (2+4i)(9) \end{bmatrix}$$

$$= \begin{bmatrix} 8+16i & -4-8i & 0 \\ 0 & 2+4i & 4+8i \\ 14+28i & 10+20i & 18+36i \end{bmatrix}$$

Matrix Multiplication

Matrix multiplication, where we multiply two matrices together, is a bit more complicated. Since it is so complicated, we begin by considering a special kind of multiplication, multiplying a *column vector* by a *row vector*.

COLUMN VECTOR

A column vector is an $n \times 1$ matrix, that is a single column with n entries. For example, let A, B, C be column vectors with two, three, and four elements, respectively.

$$A = \begin{bmatrix} -2 \\ 3 \end{bmatrix}, \qquad B = \begin{bmatrix} 9 \\ -7 \\ 11 \end{bmatrix}, \qquad C = \begin{bmatrix} 0 \\ 2 \\ -3 \\ 1 \end{bmatrix}$$

ROW VECTOR

A row vector is a $1 \times n$ matrix, or a matrix with a single row containing n elements. As an example we let D, E, F be three row vectors with two, three, and four elements, respectively.

$$A = \begin{bmatrix} 4 & -1 \end{bmatrix}, \qquad B = \begin{bmatrix} 0 & 7 & 1 \end{bmatrix}, \qquad C = \begin{bmatrix} 3 & -1 & 2 & 4 \end{bmatrix}$$

MULTIPLICATION OF A COLUMN VECTOR AND ROW VECTOR

Let $A = [a_i]$ and $B = [b_i]$ represent row and column vectors, respectively, each containing n elements. Then their product is given by

$$AB = \begin{bmatrix} a_1 & a_2 & \cdots & a_n \end{bmatrix} \begin{bmatrix} b_1 \\ b_2 \\ \vdots \\ b_n \end{bmatrix} = a_1b_1 + a_2b_2 + \cdots + a_nb_n$$

Notice that the matrix product of a row vector and a column vector is a *number*. The matrix product between a row vector and a column vector is valid only if both have the same number of elements.

EXAMPLE 2-3
Suppose that

$$A = \begin{bmatrix} 2 & 4 & -7 \end{bmatrix}, \qquad B = \begin{bmatrix} -1 \\ 2 \\ 1 \end{bmatrix}$$

Compute the product, AB.

SOLUTION 2-3
Using the formula above, we find

$$AB = \begin{bmatrix} 2 & 4 & -7 \end{bmatrix} \begin{bmatrix} -1 \\ 2 \\ 1 \end{bmatrix} = (2)(-1) + (4)(2) + (-7)(1) = -2 + 8 - 7$$

$$= -1$$

MULTIPLICATION OF MATRICES IN GENERAL

Now that we have seen how to handle the special case of matrix multiplication of a row vector and a column vector, we can tackle matrix multiplication for matrices of arbitrary dimension. First we define $A = [a_{ij}]$ as an $m \times p$ matrix and $B = [b_{ij}]$ as a $p \times n$ matrix. If we define a third matrix C such that $C=AB$, then the components of C are calculated from

$$c_{ij} = a_{i1}b_{1j} + a_{i2}b_{2j} + \cdots + a_{ip}b_{pj} = \sum_{k=1}^{p} a_{ik}b_{kj}$$

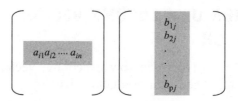

Fig. 2-1. Matrix multiplication is the product of the ith row of A and the jth column of B.

In fact the component c_{ij} is formed by multiplying the ith row of A by the jth column of B. Matrix multiplication is valid only if the number of columns of A is the same as the number of rows of B. The matrix C will have m rows and n columns (see Fig. 2-1).

EXAMPLE 2-4
Compute AB for the matrices

$$A = \begin{bmatrix} 4 & 0 & -1 \\ 1 & 2 & 3 \end{bmatrix}, \qquad B = \begin{bmatrix} 3 & 2 & -1 \\ -1 & +1 & -2 \\ 4 & -1 & 0 \end{bmatrix}$$

SOLUTION 2-4
First we check to see if the number of columns of A is the same as the number of rows of B. The matrix A has three columns and the matrix B has three rows, and so it is possible to calculate AB. Notice that the number of columns of B is 3 and the number of rows of A is 2, and so we could not calculate the product BA.

So, proceeding, the matrix $C = AB$ will have two rows and three columns, because A has two rows and B has three columns. The first component of the matrix is found by multiplying the first row of A by the first column of B. To illustrate the process, we show only the row and column of matrix A and B that are involved in each calculation. We have

$$AB = \begin{bmatrix} 4 & 0 & -1 \end{bmatrix} \begin{bmatrix} 3 \\ -1 \\ 4 \end{bmatrix} = \begin{bmatrix} (4)(3) + (0)(-1) + (-1)(4) \end{bmatrix}$$

$$= \begin{bmatrix} 8 \end{bmatrix}$$

Next, to find the element at row 1, column 2, we multiply the first row of A by the second column of B:

$$AB = \begin{bmatrix} 4 & 0 & -1 \end{bmatrix} \begin{bmatrix} 2 \\ 1 \\ -1 \end{bmatrix} = \begin{bmatrix} 8 & (4)(2) + (0)(1) + (-1)(-1) \end{bmatrix}$$

$$= \begin{bmatrix} 8 & 9 \end{bmatrix}$$

To find the element that belongs in the first row and third column of C, we multiply the first row of A by the third column of B:

$$AB = \begin{bmatrix} 4 & 0 & -1 \end{bmatrix} \begin{bmatrix} -1 \\ -2 \\ 0 \end{bmatrix} = \begin{bmatrix} 8 & 9 & (4)(-1) + (0)(-2) + (-1)(0) \end{bmatrix}$$

$$= \begin{bmatrix} 8 & 9 & -4 \end{bmatrix}$$

To fill in the second row of matrix C, we proceed as we did above but this time we use the second row of A to perform each multiplication. The first element of the second row of C is found by multiplying the second row of A by the first column of B:

$$AB = \begin{bmatrix} 1 & 2 & 3 \end{bmatrix} \begin{bmatrix} 3 \\ -1 \\ 4 \end{bmatrix} = \begin{bmatrix} 8 & 9 & -4 \\ (1)(3) + (2)(-1) + (3)(4) & & \end{bmatrix}$$

$$= \begin{bmatrix} 8 & 9 & -4 \\ 13 & & \end{bmatrix}$$

The element positioned at the second row and second column of C is found by multiplying the second row of A by the second column of B:

$$AB = \begin{bmatrix} 1 & 2 & 3 \end{bmatrix} \begin{bmatrix} 2 \\ 1 \\ -1 \end{bmatrix} = \begin{bmatrix} 8 & 9 & -4 \\ 13 & (1)(2) + (2)(1) + (3)(-1) & \end{bmatrix}$$

$$= \begin{bmatrix} 8 & 9 & -4 \\ 13 & 1 & \end{bmatrix}$$

Finally, to compute the element at the second row and third column of C, we multiply the second row of A by the third column of B:

$$AB = \begin{bmatrix} 1 & 2 & 3 \end{bmatrix} \begin{bmatrix} -1 \\ -2 \\ 0 \end{bmatrix} = \begin{bmatrix} 8 & 9 & -4 \\ 13 & 1 & (1)(-1) + (2)(-2) + (3)(0) \end{bmatrix}$$

$$= \begin{bmatrix} 8 & 9 & -4 \\ 13 & 1 & -5 \end{bmatrix}$$

In summary, we have found

$$C = AB = \begin{bmatrix} 4 & 0 & -1 \\ 1 & 2 & 3 \end{bmatrix} \begin{bmatrix} 3 & 2 & -1 \\ -1 & -1 & -2 \\ 4 & 1 & 0 \end{bmatrix} = \begin{bmatrix} 8 & 9 & -4 \\ 13 & 1 & -5 \end{bmatrix}$$

Square Matrices

A *square matrix* is a matrix that has the same number of rows and columns. We denote an $n \times n$ square matrix as a matrix of order n. While in the previous example we found that we could compute AB but it was not possible to compute BA, in many cases we work with square matrices where it is always possible to compute both multiplications. However, note that these products may not be equal.

COMMUTING MATRICES

Let $A = \begin{bmatrix} a_{ij} \end{bmatrix}$ and $B = \begin{bmatrix} b_{ij} \end{bmatrix}$ be two square $n \times n$ matrices. We say that the matrices *commute* if

$$AB = BA$$

If $AB \neq BA$, we say that the matrices do not commute.

THE COMMUTATOR

The *commutator* of two matrices A and B is denoted by $[A, B]$ and is computed using

$$[A, B] = AB - BA$$

The commutator of two matrices is a *matrix*.

EXAMPLE 2-5

Consider the following matrices:

$$A = \begin{bmatrix} 2 & -1 \\ 4 & 3 \end{bmatrix}, \qquad B = \begin{bmatrix} 1 & -4 \\ 4 & -1 \end{bmatrix}$$

Do these matrices commute?

SOLUTION 2-5

First we compute the matrix product AB:

$$AB = \begin{bmatrix} 2 & -1 \\ 4 & 3 \end{bmatrix} \begin{bmatrix} 1 & -4 \\ 4 & -1 \end{bmatrix} = \begin{bmatrix} (2)(1)+(-1)(4) & (2)(-4)+(-1)(-1) \\ (4)(1)+(3)(4) & (4)(-4)+(3)(-1) \end{bmatrix}$$

$$= \begin{bmatrix} -2 & -7 \\ 16 & -19 \end{bmatrix}$$

Remember, the element at the ith row and jth column of the matrix formed by the product is calculated by multiplying the ith row of A by the jth column of B. Now we compute the matrix product BA:

$$BA = \begin{bmatrix} 1 & -4 \\ 4 & -1 \end{bmatrix} \begin{bmatrix} 2 & -1 \\ 4 & 3 \end{bmatrix} = \begin{bmatrix} (1)(2)+(-4)(4) & (1)(-1)+(-4)(3) \\ (4)(2)+(-1)(4) & (4)(-1)+(-1)(3) \end{bmatrix}$$

$$= \begin{bmatrix} -14 & -13 \\ 4 & -7 \end{bmatrix}$$

We notice immediately that $AB \neq BA$ and so the matrices do not commute. The commutator is found to be

$$[A, B] = AB - BA = \begin{bmatrix} -2 & -7 \\ 16 & -19 \end{bmatrix} - \begin{bmatrix} -14 & -13 \\ 4 & -7 \end{bmatrix}$$

$$= \begin{bmatrix} -2-(-14) & -7-(-13) \\ 16-4 & -19-(-7) \end{bmatrix} = \begin{bmatrix} 12 & 6 \\ 12 & -12 \end{bmatrix}$$

The commutator is another matrix with the same number of rows and columns as A and B.

EXAMPLE 2-6

Let

$$A = \begin{bmatrix} 1 & -x \\ x & 1 \end{bmatrix}, \qquad B = \begin{bmatrix} 2 & y \\ 1 & -y \end{bmatrix}$$

Find x and y such that the commutator of these two matrices is zero, i.e., $AB = BA$.

SOLUTION 2-6

We compute AB:

$$AB = \begin{bmatrix} 1 & -x \\ x & 1 \end{bmatrix}\begin{bmatrix} 2 & y \\ 1 & -y \end{bmatrix} = \begin{bmatrix} (1)(2)+(-x)(1) & (1)(y)+(-x)(-y) \\ (x)(2)+(1)(1) & (x)(y)+(1)(-y) \end{bmatrix}$$

$$= \begin{bmatrix} 2-x & y+xy \\ 2x+1 & xy-y \end{bmatrix}$$

For BA we find

$$BA = \begin{bmatrix} 2 & y \\ 1 & -y \end{bmatrix}\begin{bmatrix} 1 & -x \\ x & 1 \end{bmatrix} = \begin{bmatrix} (2)(1)+(y)(x) & (2)(-x)+(y)(1) \\ (1)(1)+(-y)(x) & (1)(-x)+(-y)(1) \end{bmatrix}$$

$$= \begin{bmatrix} 2+xy & -2x+y \\ 1-xy & -x-y \end{bmatrix}$$

For these to be equal we must have

$$\begin{bmatrix} 2-x & y+xy \\ 2x+1 & xy-y \end{bmatrix} = \begin{bmatrix} 2+xy & -2x+y \\ 1-xy & -x-y \end{bmatrix}$$

Each term that belongs to the matrix on the left-hand side must be equal to the corresponding term in the matrix on the right side. This means that

$$2-x = 2+xy$$

$$y+xy = -2x+y$$

$$2x+1 = 1-xy$$

$$xy-y = -x-y$$

We examine the first equation $2 - x = 2 + xy$. Subtracting 2 from both sides we have

$$-x = xy$$

Now divide both sides by x. This gives

$$y = -1$$

Inserting this value into the second equation

$$y + xy = -2x + y$$

we find

$$-1 - x = -2x - 1$$

Now add 1 to both sides and divide by -1, which gives

$$x = 2x$$

This equation implies that $x = 0$. You can check that these values satisfy the other two equations. Therefore, if we take $x = 0$ and $y = -1$, then the matrices are

$$A = \begin{bmatrix} 1 & 0 \\ 0 & 1 \end{bmatrix}, \qquad B = \begin{bmatrix} 2 & -1 \\ 1 & 1 \end{bmatrix}$$

The matrix A is a special matrix called the *identity* matrix.

The Identity Matrix

There exists a special matrix that plays a role analogous to the number 1 in the matrix world. This is the *identity* matrix. For any matrix A we have

$$AI = IA = A$$

where I is the identity matrix. The identity matrix is a square matrix with 1s along the diagonal and 0s everywhere else (see Fig. 2-2).

$$I_n = \begin{bmatrix} 1 & & 0 \\ & \ddots & \\ 0 & & 1 \end{bmatrix}$$

Fig. 2-2. A general representation of the identity matrix.

The 2×2 identity matrix is given by

$$I_2 = \begin{bmatrix} 1 & 0 \\ 0 & 1 \end{bmatrix}$$

and the 3×3 identity matrix is given by

$$\begin{bmatrix} 1 & 0 & 0 \\ 0 & 1 & 0 \\ 0 & 0 & 1 \end{bmatrix}$$

Higher order identity matrices are defined similarly.

EXAMPLE 2-7
Verify that the 2×2 identity matrix satisfies $AI = IA = A$ for the matrix A defined by

$$A = \begin{bmatrix} 2 & 8 \\ -7 & 4 \end{bmatrix}$$

SOLUTION 2-7
We have

$$I_2 = \begin{bmatrix} 1 & 0 \\ 0 & 1 \end{bmatrix}$$

And so

$$AI = \begin{bmatrix} 2 & 8 \\ -7 & 4 \end{bmatrix} \begin{bmatrix} 1 & 0 \\ 0 & 1 \end{bmatrix} = \begin{bmatrix} (2)(1) + (8)(0) & (2)(0) + (8)(1) \\ (-7)(1) + (4)(0) & (-7)(0) + (4)(1) \end{bmatrix}$$

$$= \begin{bmatrix} 2 & 8 \\ -7 & 4 \end{bmatrix} = A$$

Performing the multiplication in the opposite order, we obtain

$$IA = \begin{bmatrix} 1 & 0 \\ 0 & 1 \end{bmatrix} \begin{bmatrix} 2 & 8 \\ -7 & 4 \end{bmatrix} = \begin{bmatrix} (1)(2)+(0)(-7) & (1)(8)+(0)(4) \\ (0)(2)+(1)(-7) & (0)(8)+(1)(4) \end{bmatrix}$$

$$= \begin{bmatrix} 2 & 8 \\ -7 & 4 \end{bmatrix} = A$$

The Transpose Operation

The transpose of a matrix is found by exchanging the rows and columns of the matrix (see Fig. 2-3) and denoted by

$$\text{transpose}\,(A) = A^T$$

This is best demonstrated by an example. Let

$$A = \begin{bmatrix} 1 & 2 & 3 \\ 4 & 5 & 6 \end{bmatrix}$$

Then we have

$$A^T = \begin{bmatrix} 1 & 4 \\ 2 & 5 \\ 3 & 6 \end{bmatrix}$$

Notice that if A is an $m \times n$ matrix, then the transpose A^T is an $n \times m$ matrix. We often compute the transpose of a square matrix. If

$$A = \begin{bmatrix} 0 & 1 \\ 2 & 3 \end{bmatrix}$$

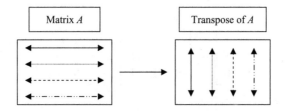

Fig. 2-3. A schematic representation of the transpose operation. The rows of a matrix become the columns of the transpose matrix.

Then the transpose is

$$A^T = \begin{bmatrix} 0 & 2 \\ 1 & 3 \end{bmatrix}$$

We can take the transpose of any sized matrix. For example,

$$B = \begin{bmatrix} -1 & 2 & 0 \\ 0 & 1 & 4 \\ 5 & 5 & 6 \end{bmatrix}, \qquad B^T = \begin{bmatrix} -1 & 0 & 5 \\ 2 & 1 & 5 \\ 0 & 4 & 6 \end{bmatrix}$$

The transpose operation satisfies several properties:

- $(A + B)^T = A^T + B^T$
- $(\alpha A)^T = \alpha A^T$, where α is a scalar
- $(A^T)^T = A$
- $(AB)^T = B^T A^T$

EXAMPLE 2-8
Let

$$A = \begin{bmatrix} 1 & 0 & 1 \\ -2 & 1 & 3 \\ 4 & 1 & 0 \end{bmatrix}, \qquad B = \begin{bmatrix} 2 & 2 & 1 \\ 1 & 3 & 1 \\ 4 & 1 & 1 \end{bmatrix}$$

Show that these matrices satisfy $(A + B)^T = A^T + B^T$ and $(AB)^T = B^T A^T$.

SOLUTION 2-8
We begin by adding the matrices

$$A + B = \begin{bmatrix} 1 & 0 & 1 \\ -2 & 1 & 3 \\ 4 & 1 & 0 \end{bmatrix} + \begin{bmatrix} 2 & 2 & 1 \\ 1 & 3 & 1 \\ 4 & 1 & 1 \end{bmatrix} = \begin{bmatrix} 1+2 & 0+2 & 1+1 \\ -2+1 & 1+3 & 3+1 \\ 4+4 & 1+1 & 0+1 \end{bmatrix}$$

$$= \begin{bmatrix} 3 & 2 & 2 \\ -1 & 4 & 4 \\ 8 & 2 & 1 \end{bmatrix}$$

To compute the transpose, we exchange rows and columns. The transpose of the sum is found to be

$$(A + B)^T = \begin{bmatrix} 3 & 2 & 2 \\ -1 & 4 & 4 \\ 8 & 2 & 1 \end{bmatrix}^T = \begin{bmatrix} 3 & -1 & 8 \\ 2 & 4 & 2 \\ 2 & 4 & 1 \end{bmatrix}$$

Now we compute the transpose of each individual matrix:

$$A^T = \begin{bmatrix} 1 & 0 & 1 \\ -2 & 1 & 3 \\ 4 & 1 & 0 \end{bmatrix}^T = \begin{bmatrix} 1 & -2 & 4 \\ 0 & 1 & 1 \\ 1 & 3 & 0 \end{bmatrix},$$

$$B^T = \begin{bmatrix} 2 & 2 & 1 \\ 1 & 3 & 1 \\ 4 & 1 & 1 \end{bmatrix}^T = \begin{bmatrix} 2 & 1 & 4 \\ 2 & 3 & 1 \\ 1 & 1 & 1 \end{bmatrix}$$

We add the transpose of each matrix together and we obtain

$$A^T + B^T = \begin{bmatrix} 1 & -2 & 4 \\ 0 & 1 & 1 \\ 1 & 3 & 0 \end{bmatrix} + \begin{bmatrix} 2 & 1 & 4 \\ 2 & 3 & 1 \\ 1 & 1 & 1 \end{bmatrix} = \begin{bmatrix} 1+2 & -2+1 & 4+4 \\ 0+2 & 1+3 & 1+1 \\ 1+1 & 3+1 & 0+1 \end{bmatrix}$$

$$= \begin{bmatrix} 3 & -1 & 8 \\ 2 & 4 & 2 \\ 2 & 4 & 1 \end{bmatrix} = (A + B)^T$$

Now we show that $(AB)^T = B^T A^T$. First we compute the product of the two matrices. Remember, the element at c_{ij} is found by multiplying the ith row of A by the jth column of B. We find

$$AB = \begin{bmatrix} 1 & 0 & 1 \\ -2 & 1 & 3 \\ 4 & 1 & 0 \end{bmatrix} \begin{bmatrix} 2 & 2 & 1 \\ 1 & 3 & 1 \\ 4 & 1 & 1 \end{bmatrix}$$

$$= \begin{bmatrix} (1)(2)+(0)(1)+(1)(4) & (1)(2)+(0)(3)+(1)(1) & (1)(1)+(0)(1)+(1)(1) \\ (-2)(2)+(1)(1)+(3)(4) & (-2)(2)+(1)(3)+(3)(1) & (-2)(1)+(1)(1)+(3)(1) \\ (4)(2)+(1)(1)+(0)(4) & (4)(2)+(1)(3)+(0)(1) & (4)(1)+(1)(1)+(0)(1) \end{bmatrix}$$

$$= \begin{bmatrix} 6 & 3 & 2 \\ 9 & 2 & 2 \\ 9 & 11 & 5 \end{bmatrix}$$

The transpose is found by exchanging the rows and columns

$$(AB)^T = \begin{bmatrix} 6 & 3 & 2 \\ 9 & 2 & 2 \\ 9 & 11 & 5 \end{bmatrix}^T = \begin{bmatrix} 6 & 9 & 9 \\ 3 & 2 & 11 \\ 2 & 2 & 5 \end{bmatrix}$$

Using the A^T and B^T matrices that we calculated above

$$B^T A^T = \begin{bmatrix} 2 & 1 & 4 \\ 2 & 3 & 1 \\ 1 & 1 & 1 \end{bmatrix} \begin{bmatrix} 1 & -2 & 4 \\ 0 & 1 & 1 \\ 1 & 3 & 0 \end{bmatrix}$$

$$= \begin{bmatrix} (2)(1)+(1)(0)+(4)(1) & (2)(-2)+(1)(1)+(4)(3) & (2)(4)+(1)(1)+(4)(0) \\ (2)(1)+(3)(0)+(1)(1) & (2)(-2)+(3)(1)+(1)(3) & (2)(4)+(3)(1)+(1)(0) \\ (1)(1)+(1)(0)+(1)(1) & (1)(-2)+(1)(1)+(1)(3) & (1)(4)+(1)(1)+(1)(0) \end{bmatrix}$$

$$= \begin{bmatrix} 6 & 9 & 9 \\ 3 & 2 & 11 \\ 2 & 2 & 5 \end{bmatrix} = (AB)^T$$

EXAMPLE 2-9
Prove that

$$(A + B)^T = A^T + B^T$$

SOLUTION 2-9
The first thing we note is that in order to add the matrices A and B together, they must both have the same number of rows and columns. In other words, if A is an $m \times n$ matrix, then so is B, and $A + B$ is an $m \times n$ matrix as well. This means that $(A + B)^T$ is an $n \times m$ matrix.

On the right-hand side, if A and B are $m \times n$ matrices, then clearly A^T and B^T are $n \times m$ matrices.

To verify the property, it is sufficient to show that the (i, j) entry on both sides of the equation are the same. First we examine the right-hand side. If the (i, j) entry of A is denoted by a_{ij}, then the (i, j) entry of the transpose is given by changing the row and column, i.e., the (i, j) entry of A^T is given by a_{ji}. Similarly

the (i, j) of B^T is b_{ji}. We summarize this by writing

$$\left(A^T + B^T\right)_{ij} = a_{ji} + b_{ji}$$

On the left-hand side, the (i, j) entry of $C = A + B$ is

$$c_{ij} = a_{ij} + b_{ij}$$

The (i, j) element of C^T is also found by swapping row and column indices, and so the (i, j) entry of $(A + B)^T$ is

$$(A + B)^T_{ij} = a_{ji} + b_{ji}$$

The (i, j) element of both sides are equal; therefore, the matrices are the same.

The Hermitian Conjugate

We now extend the transpose operation to the Hermitian conjugate, which is written as

$$A^\dagger$$

(read "A dagger"). The Hermitian conjugate applies to matrices with complex elements and is a two-step operation (see Fig. 2-4):

- Take the transpose of the matrix.
- Take the complex conjugate of the elements.

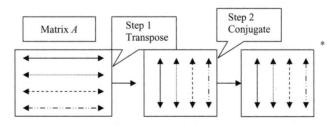

Fig. 2-4. A schematic representation of finding the Hermitian conjugate of a matrix. Take the transpose, turning rows into columns, and then compute the complex conjugate of each element.

EXAMPLE 2-10
Find $A^\dagger$ for

$$A = \begin{bmatrix} -9 & 2i & 0 \\ 0 & 4i & 7 \\ 1+2i & 3 & 0 \end{bmatrix}$$

SOLUTION 2-10
Step 1, we take the transpose of the matrix:

$$A^T = \begin{bmatrix} -9 & 2i & 0 \\ 0 & 4i & 7 \\ 1+2i & 3 & 0 \end{bmatrix}^T = \begin{bmatrix} -9 & 0 & 1+2i \\ 2i & 4i & 3 \\ 0 & 7 & 0 \end{bmatrix}$$

Now we apply step 2 by taking the complex conjugate of each element. This means that we let $i \to -i$. We find

$$A^\dagger = \left(A^T\right)^* = \begin{bmatrix} -9 & 0 & 1+2i \\ 2i & 4i & 3 \\ 0 & 7 & 0 \end{bmatrix}^* = \begin{bmatrix} -9 & 0 & 1-2i \\ -2i & -4i & 3 \\ 0 & 7 & 0 \end{bmatrix}$$

Trace

The *trace* of a square $n \times n$ matrix is found by summing the diagonal elements. We denote the trace of a matrix A by writing $\text{tr}(A)$. If the matrix elements of A are given by a_{ij} then

$$tr(A) = a_{11} + a_{22} + \cdots + a_{nn} = \sum_{i=1}^{n} a_{ii}$$

The trace operation has the following properties:

- $\text{tr}(\alpha A) = \alpha \, \text{tr}(A)$
- $\text{tr}(A + B) = \text{tr}(A) + \text{tr}(B)$
- $\text{tr}(AB) = \text{tr}(BA)$

EXAMPLE 2-11
Find the trace of the matrix

$$B = \begin{bmatrix} -1 & 7 & 0 & 1 \\ 0 & 5 & 2 & 1 \\ 1 & 0 & 1 & 2 \\ 0 & 4 & 4 & 8 \end{bmatrix}$$

SOLUTION 2-11

The trace of a matrix is the sum of the diagonal elements, and so

$$\text{tr}(B) = -1 + 5 + 1 + 8 = 13$$

EXAMPLE 2-12

Verify that $\text{tr}(\alpha A) = \alpha \, \text{tr}(A)$ for $\alpha = 3$ and

$$A = \begin{bmatrix} 2 & -1 \\ -1 & 7 \end{bmatrix}$$

SOLUTION 2-12

The scalar multiplication of A is given by

$$\alpha A = (3) \begin{bmatrix} 2 & -1 \\ -1 & 7 \end{bmatrix} = \begin{bmatrix} (3)(2) & (3)(-1) \\ (3)(-1) & (3)(7) \end{bmatrix} = \begin{bmatrix} 6 & -3 \\ -3 & 21 \end{bmatrix}$$

The trace is the sum of the diagonal elements

$$\text{tr}(\alpha A) = 6 + 21 = 27$$

Now the trace of A is

$$\text{tr}(A) = \text{tr} \begin{bmatrix} 2 & -1 \\ -1 & 7 \end{bmatrix} = 2 + 7 = 9$$

Therefore we find that

$$\alpha \, \text{tr}(A) = 3(9) = 27 = \text{tr}(\alpha A)$$

EXAMPLE 2-13

Prove that $\text{tr}(A + B) = \text{tr}(A) + \text{tr}(B)$.

SOLUTION 2-13
On the left side we have

$$\text{tr}\,(A+B) = \sum_{i=1}^{n} a_{ii} + b_{ii}$$

On the right we have

$$\text{tr}\,(A) + \text{tr}\,(B) = \sum_{i=1}^{n} a_{ii} + \sum_{i=1}^{n} b_{ii}$$

We can combine these sums into a single sum, proving the result

$$\sum_{i=1}^{n} a_{ii} + \sum_{i=1}^{n} b_{ii} = \sum_{i=1}^{n} a_{ii} + b_{ii} = \text{tr}\,(A+B)$$

The Inverse Matrix

The inverse of an $n \times n$ square matrix A is denoted by A^{-1} and has the property
that

$$A A^{-1} = A^{-1} A = I$$

The components of the inverse of a matrix can be found by brute force multiplication. Later we will explore a more sophisticated way to obtain the inverse using determinants. A matrix with an inverse is called *nonsingular*.

EXAMPLE 2-14
Let

$$A = \begin{bmatrix} 2 & 3 \\ -1 & 4 \end{bmatrix}$$

and find its inverse.

SOLUTION 2-14
We denote the inverse matrix by

$$A^{-1} = \begin{bmatrix} a & b \\ c & d \end{bmatrix}$$

We compute AA^{-1}:

$$AA^{-1} = \begin{bmatrix} 2 & 3 \\ -1 & 4 \end{bmatrix} \begin{bmatrix} a & b \\ c & d \end{bmatrix} = \begin{bmatrix} 2a + 3c & 2b + 3d \\ -a + 4c & -b + 4d \end{bmatrix}$$

The equation $AA^{-1} = I$ means that

$$\begin{bmatrix} 2a + 3c & 2b + 3d \\ -a + 4c & -b + 4d \end{bmatrix} = \begin{bmatrix} 1 & 0 \\ 0 & 1 \end{bmatrix}$$

Equating element by element gives four equations for four unknowns:

$$2a + 3c = 1$$
$$2b + 3d = 0 \quad \Rightarrow d = -\frac{2}{3}b$$
$$-a + 4c = 0 \quad \Rightarrow a = 4c$$
$$-b + 4d = 1$$

Substitution of $a = 4c$ into the first equation gives

$$2a + 3c = 2(4c) + 3c = 8c + 3c = 11c = 1$$
$$\Rightarrow c = \frac{1}{11}, \quad a = \frac{4}{11}$$

Now we substitute $d = -\frac{2}{3}b$ into the last equation, which gives

$$-b + 4d = -b - \frac{8}{3}b = -\frac{11}{3}b = 1$$
$$\Rightarrow b = -\frac{3}{11}, \quad d = -\frac{2}{3}b = \frac{2}{11}$$

And so the inverse is

$$A^{-1} = \begin{bmatrix} \dfrac{4}{11} & -\dfrac{3}{11} \\[2mm] \dfrac{1}{11} & \dfrac{2}{11} \end{bmatrix}$$

We double-check the result:

$$AA^{-1} = \begin{bmatrix} 2 & 3 \\ -1 & 4 \end{bmatrix} \begin{bmatrix} \dfrac{4}{11} & -\dfrac{3}{11} \\ \dfrac{1}{11} & \dfrac{2}{11} \end{bmatrix} = \begin{bmatrix} \dfrac{8}{11} + \dfrac{3}{11} & -\dfrac{6}{11} + \dfrac{6}{11} \\ -\dfrac{4}{11} + \dfrac{4}{11} & \dfrac{3}{11} + \dfrac{8}{11} \end{bmatrix}$$

$$= \begin{bmatrix} \dfrac{11}{11} & 0 \\ 0 & \dfrac{11}{11} \end{bmatrix} = \begin{bmatrix} 1 & 0 \\ 0 & 1 \end{bmatrix}$$

PROPERTIES OF THE INVERSE

The inverse operation satisfies

- $\left(A^{-1}\right)^{-1} = A$
- $(\alpha A)^{-1} = \dfrac{1}{\alpha} A^{-1}$
- $\left(A^{-1}\right)^{T} = \left(A^{T}\right)^{-1}$
- $(AB)^{-1} = B^{-1} A^{-1}$

EXAMPLE 2-15
Prove that if A and B are invertible, then $(AB)^{-1} = B^{-1} A^{-1}$.

SOLUTION 2-15
If A and B are invertible, we know that

$$AA^{-1} = A^{-1}A = I$$
$$BB^{-1} = B^{-1}B = I$$

Now we have

$$(AB)\, B^{-1} A^{-1} = A\left(BB^{-1}\right) A^{-1} = A\,(I)\,A^{-1} = AA^{-1} = I$$

Multiplying these terms in the opposite order, we have

$$B^{-1} A^{-1}\,(AB) = B^{-1}\left(A^{-1}A\right) B = B^{-1}\,(I)\,B = BB^{-1} = I$$

Since both of these relations are true, then

$$(AB)^{-1} = B^{-1}A^{-1}$$

Note that an $n \times n$ linear system $Ax = b$ has solution $x = A^{-1}b$ if the matrix A is nonsingular.

EXAMPLE 2-16
Solve the linear system

$$2x + 3y = 4$$
$$2x + y = -1$$

by finding a solution to $Ax = b$.

SOLUTION 2-16
We write the system as

$$\begin{bmatrix} 2 & 3 \\ 2 & 1 \end{bmatrix} \begin{bmatrix} x \\ y \end{bmatrix} = \begin{bmatrix} 4 \\ -1 \end{bmatrix}$$

The inverse of the matrix

$$A = \begin{bmatrix} 2 & 3 \\ 2 & 1 \end{bmatrix}$$

is

$$A^{-1} = \begin{bmatrix} -\dfrac{1}{4} & \dfrac{3}{4} \\ \dfrac{1}{2} & -\dfrac{1}{2} \end{bmatrix}$$

(verify this). The solution is

$$\begin{bmatrix} x \\ y \end{bmatrix} = \begin{bmatrix} -\dfrac{1}{4} & \dfrac{3}{4} \\ \dfrac{1}{2} & -\dfrac{1}{2} \end{bmatrix} \begin{bmatrix} 4 \\ -1 \end{bmatrix}$$

Carrying out the matrix multiplication on the right side, we find

$$
\begin{bmatrix} -\dfrac{1}{4} & \dfrac{3}{4} \\[2mm] \dfrac{1}{2} & -\dfrac{1}{2} \end{bmatrix} \begin{bmatrix} 4 \\ -1 \end{bmatrix} = \begin{bmatrix} -\dfrac{7}{4} \\[2mm] \dfrac{5}{2} \end{bmatrix} = \begin{bmatrix} x \\ y \end{bmatrix}
$$

$$
\Rightarrow x = -\frac{7}{4}, \quad y = \frac{5}{2}
$$

We verify that these values satisfy the equations. The first equation is

$$2x + 3y = 4$$

$$
\Rightarrow 2\left(-\frac{7}{4}\right) + 3\left(\frac{5}{2}\right) = -\frac{14}{4} + \frac{15}{2} = -\frac{14}{4} + \frac{30}{4} = \frac{16}{4} = 4
$$

For the second equation we have

$$2x + y = -1$$

$$
\Rightarrow 2\left(-\frac{7}{4}\right) + \left(\frac{5}{2}\right) = -\frac{14}{4} + \frac{5}{2} = -\frac{14}{4} + \frac{10}{4} = -\frac{4}{4} = -1
$$

While this small system is extremely simple and could be solved by hand, the technique illustrated is very valuable for solving large systems of equations.

Quiz

1. For the matrices given by

$$
A = \begin{bmatrix} -2 & 1 & 0 \\ 9 & 4 & -3 \\ 2 & 1 & 0 \end{bmatrix}, \qquad B = \begin{bmatrix} 1 & -1 & 0 \\ 2 & 4 & 5 \\ 9 & 8 & 1 \end{bmatrix}
$$

calculate
(a) $A + B$
(b) αA for $\alpha = 2$
(c) AB

2. Find the matrix product for

$$A = \begin{bmatrix} 2 & -1 & 4 \end{bmatrix}, \qquad B = \begin{bmatrix} 1 \\ 7 \\ 1 \end{bmatrix}$$

3. Find the commutator of the matrices

$$A = \begin{bmatrix} 2 & 2 & -1 \\ 4 & 0 & -1 \\ 3 & 1 & 5 \end{bmatrix}, \qquad B = \begin{bmatrix} 1 & 3 & 1 \\ 5 & 1 & 0 \\ 3 & 0 & 0 \end{bmatrix}$$

4. Can you find the value of x such that $AB = BA$ for

$$A = \begin{bmatrix} x & 1 \\ 2 & x \end{bmatrix}, \qquad B = \begin{bmatrix} -1 & 0 \\ 1 & 4 \end{bmatrix}$$

5. Find the trace of the matrix

$$A = \begin{bmatrix} 8 & 0 & 0 & -1 \\ 7 & 9 & 1 & 0 \\ 2 & 0 & 0 & 1 \\ 9 & -8 & 17 & -1 \end{bmatrix}$$

6. Prove that $\operatorname{tr}(\alpha A) = \alpha \operatorname{tr}(A)$.
7. For the matrices

$$A = \begin{bmatrix} i & 7 \\ 3+i & 2-i \end{bmatrix}, \qquad B = \begin{bmatrix} 9 & 6+3i \\ 1+i & 4 \end{bmatrix}$$

    (a) calculate the commutator $[A, B] = AB - BA$
    (b) find $\operatorname{tr}(A)$, $\operatorname{tr}(B)$

8. For the matrices

$$A = \begin{bmatrix} 1 & -1 & 5 \\ 0 & 4 & 0 \\ 1 & 1 & -2 \end{bmatrix}, \qquad B = \begin{bmatrix} 9 & -1 & 0 \\ 8 & 8 & 4 \\ 16 & 0 & 1 \end{bmatrix}$$

    (a) find A^T and B^T
    (b) show that $(A + B)^T = A^T + B^T$

9. Does the matrix

$$A = \begin{bmatrix} 7 & -1 & 0 \\ 0 & 0 & 4 \\ 1 & 2 & -1 \end{bmatrix}$$

have an inverse? If so, find it.

10. Is there a solution to the system

$$3x - 2y + z = 9$$
$$4x + y + 3z = -1$$
$$-x + 5y + 2z = 7$$

Begin by finding the inverse of

$$A = \begin{bmatrix} 3 & -2 & 1 \\ 4 & 1 & 3 \\ -1 & 5 & 2 \end{bmatrix}$$

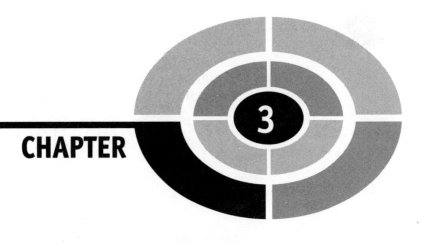

Determinants

The *determinant* of a matrix is a *number* that is associated with the matrix. For a matrix A denote this number by

$$|A|$$

or by writing

$$\det |A|$$

THE DETERMINANT OF A SECOND-ORDER MATRIX

The determinant of second-order square matrix

$$A = \begin{bmatrix} a & b \\ c & d \end{bmatrix}$$

is given by the number

$$ad - bc$$

EXAMPLE 3-1
Find the determinant of

$$A = \begin{bmatrix} 2 & 8 \\ 1 & 4 \end{bmatrix}$$

SOLUTION 3-1

$$\det |A| = \det \left| \begin{bmatrix} 2 & 8 \\ 1 & 4 \end{bmatrix} \right| = (2)(4) - (1)(8) = 8 - 8 = 0$$

EXAMPLE 3-2
Find the determinant of

$$A = \begin{bmatrix} 7 & 3 \\ 1 & 4 \end{bmatrix}$$

SOLUTION 3-2

$$\det |A| = \det \left| \begin{bmatrix} 7 & 3 \\ 1 & 4 \end{bmatrix} \right| = (7)(4) - (1)(3) = 28 - 3 = 25$$

The determinant can be calculated for matrices of complex numbers as well.

EXAMPLE 3-3
Find the determinant of

$$B = \begin{bmatrix} -2i & 1 \\ 4 & 6 + 2i \end{bmatrix}$$

SOLUTION 3-3
Recalling that $i^2 = -1$, we obtain

$$\det |B| = \det \left| \begin{bmatrix} -2i & 1 \\ 4 & 6 + 2i \end{bmatrix} \right| = (-2i)(6 + 2i) - (1)(4)$$

$$= -12i + 4 - 4$$

$$= -12i$$

The Determinant of a Third-Order Matrix

Let a 3×3 matrix A be given by

$$A = \begin{bmatrix} a_1 & a_2 & a_3 \\ b_1 & b_2 & b_3 \\ c_1 & c_2 & c_3 \end{bmatrix}$$

The determinant of A is given by (see Fig. 3-1)

$$\det |A| = a_1 \det \begin{vmatrix} b_2 & b_3 \\ c_2 & c_3 \end{vmatrix} - a_2 \det \begin{vmatrix} b_1 & b_3 \\ c_1 & c_3 \end{vmatrix} + a_3 \det \begin{vmatrix} b_1 & b_2 \\ c_1 & c_2 \end{vmatrix}$$

$$= a_1 (b_2 c_3 - c_2 b_3) - a_2 (b_1 c_3 - c_1 b_3) + a_3 (b_1 c_2 - c_1 b_2)$$

EXAMPLE 3-4

Find the determinant of

$$C = \begin{bmatrix} 4 & 0 & 2 \\ 1 & -2 & 5 \\ 1 & 0 & 1 \end{bmatrix}$$

SOLUTION 3-4

$$\det |C| = \det \begin{bmatrix} 4 & 0 & 2 \\ 1 & -2 & 5 \\ 1 & 0 & 1 \end{bmatrix} = (4) \det \begin{vmatrix} -2 & 5 \\ 0 & 1 \end{vmatrix} + (2) \det \begin{vmatrix} 1 & -2 \\ 1 & 0 \end{vmatrix}$$

$$\begin{pmatrix} a & b & c \\ d & e & f \\ g & h & i \end{pmatrix}$$

Fig. 3-1. When taking the determinant of a third-order matrix, the elements of the top row are coefficients of determinants formed from elements from rows 2 and 3. To find the elements used in the determinant associated with each coefficient, cross out the top row. Then cross out the column under the given coefficient. In this example the coefficient is element c, and so we cross out the third column. The leftover elements, d, e, g, h are used to construct a second-order matrix. We then take its determinant.

Now we have

$$\det \begin{vmatrix} -2 & 5 \\ 0 & 1 \end{vmatrix} = (-2)(1) - (0)(5) = -2$$

and

$$\det \begin{vmatrix} 1 & -2 \\ 1 & 0 \end{vmatrix} = (1)(0) - (-2)(1) = 2$$

Therefore we obtain

$$\det |C| = (4)(-2) + (2)(2) = -8 + 4 = -4$$

Theorems about Determinants

We now cover some important theorems involving determinants.

DETERMINANT OF A MATRIX WITH TWO IDENTICAL ROWS OR IDENTICAL COLUMNS

The determinant of a matrix with two rows or two columns that are identical is zero.

EXAMPLE 3-5
Show that the determinant of a second-order matrix with identical rows is zero.

SOLUTION 3-5
We write the matrix as

$$A = \begin{bmatrix} a & b \\ a & b \end{bmatrix}$$

The determinant is

$$\det |A| = \det \begin{bmatrix} a & b \\ a & b \end{bmatrix} = ab - ab = 0$$

SWAPPING ROWS OR COLUMNS IN A MATRIX

If we swap two rows or two columns of a matrix, the determinant changes sign.

EXAMPLE 3-6
Show that for

$$A = \begin{bmatrix} -1 & 2 \\ 4 & 8 \end{bmatrix}, \qquad B = \begin{bmatrix} 4 & 8 \\ -1 & 2 \end{bmatrix}$$

$$\det|B| = -\det|A|$$

SOLUTION 3-6
We have

$$\det|A| = \det\begin{vmatrix} -1 & 2 \\ 4 & 8 \end{vmatrix} = (-1)(8) - (2)(4) = -8 - 8 = -16$$

For the other matrix we obtain

$$\det|B| = \det\begin{vmatrix} 4 & 8 \\ -1 & 2 \end{vmatrix} = (4)(2) - (-1)(8) = 8 + 8 = 16$$

and we see that $\det|B| = -\det|A|$ is satisfied.

Cramer's Rule

Cramer's rule is a simple algorithm that allows determinants to be used to solve systems of linear equations. We examine the case of two equations with two unknowns first. Consider the system

$$ax + by = m$$
$$cx + dy = n$$

This system can be written in matrix form as

$$Ax = b$$

where we have

$$\begin{bmatrix} a & b \\ c & d \end{bmatrix}\begin{bmatrix} x \\ y \end{bmatrix} = \begin{bmatrix} m \\ n \end{bmatrix}$$

If the determinant

$$\det |A| = \det \begin{vmatrix} a & b \\ c & d \end{vmatrix} = ad - bc \neq 0$$

then Cramer's rule allows us to find a solution given by

$$x = \frac{\det \begin{vmatrix} m & c \\ n & d \end{vmatrix}}{\det \begin{vmatrix} a & b \\ c & d \end{vmatrix}}, \qquad y = \frac{\det \begin{vmatrix} a & m \\ b & n \end{vmatrix}}{\det \begin{vmatrix} a & b \\ c & d \end{vmatrix}}$$

EXAMPLE 3-7
Find a solution to the system

$$3x - y = 4$$
$$2x + y = -2$$

SOLUTION 3-7
We write the matrix A of coefficients as

$$A = \begin{bmatrix} 3 & -1 \\ 2 & 1 \end{bmatrix}$$

The determinant is

$$\det |A| = \det \begin{vmatrix} 3 & -1 \\ 2 & 1 \end{vmatrix} = (3)(1) - (2)(-1) = 3 + 2 = 5$$

Since the determinant is nonzero, we can use Cramer's rule to find a solution. We find x by substitution of the first column of A by the elements of the vector b:

$$x = \frac{\det \begin{vmatrix} 4 & -1 \\ -2 & 1 \end{vmatrix}}{\det |A|} = \frac{(4)(1) - (-2)(-1)}{5} = \frac{4 - 2}{5} = \frac{2}{5}$$

To find y, we substitute for the other column:

$$y = \frac{\det \begin{vmatrix} 3 & 4 \\ 2 & -2 \end{vmatrix}}{\det |A|} = \frac{(3)(-2) - (2)(4)}{5} = \frac{-6 - 8}{5} = -\frac{14}{5}$$

Cramer's rule can be extended to a system of three equations with three unknowns. The procedure is the same. If we have the system

$$ax + by + cz = r$$
$$dx + ey + fz = s$$
$$kx + ly + mz = t$$

(note that in this instance we are considering i as a constant), then we can write this in the form $Ax = b$ with

$$A = \begin{bmatrix} a & b & c \\ d & e & f \\ k & l & m \end{bmatrix}, \qquad X = \begin{bmatrix} x \\ y \\ z \end{bmatrix}, \qquad b = \begin{bmatrix} r \\ s \\ t \end{bmatrix}$$

Cramer's rule tells us that the solution is

$$x = \frac{\begin{vmatrix} r & b & c \\ s & e & f \\ t & l & m \end{vmatrix}}{\begin{vmatrix} a & b & c \\ d & e & f \\ k & l & m \end{vmatrix}}, \qquad y = \frac{\begin{vmatrix} a & r & c \\ d & s & f \\ k & t & m \end{vmatrix}}{\begin{vmatrix} a & b & c \\ d & e & f \\ k & l & m \end{vmatrix}}, \qquad z = \frac{\begin{vmatrix} a & b & r \\ d & e & s \\ k & l & t \end{vmatrix}}{\begin{vmatrix} a & b & c \\ d & e & f \\ k & l & m \end{vmatrix}}$$

provided that the determinant

$$\begin{vmatrix} a & b & c \\ d & e & f \\ k & l & m \end{vmatrix}$$

is nonzero. Notice that this solution gives the point of intersection of three planes (see Fig. 3-2).

EXAMPLE 3-8
Find the point of intersection of the three planes defined by

$$x + 2y - z = 4$$
$$2x - y + 3z = 3$$
$$4x + 3y - 2z = 5$$

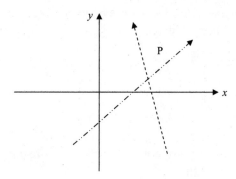

Fig. 3-2. Cramer's rule allows us to find the point P where two lines intersect.

SOLUTION 3-8
The matrix of coefficients is given by

$$A = \begin{bmatrix} 1 & 2 & -1 \\ 2 & -1 & 3 \\ 4 & 3 & -2 \end{bmatrix}$$

The determinant is

$$\det |A| = \det \begin{vmatrix} 1 & 2 & -1 \\ 2 & -1 & 3 \\ 4 & 3 & -2 \end{vmatrix}$$

$$= \det \begin{vmatrix} -1 & 3 \\ 3 & -2 \end{vmatrix} - (2) \det \begin{vmatrix} 2 & 3 \\ 4 & -2 \end{vmatrix} - \det \begin{vmatrix} 2 & -1 \\ 4 & 3 \end{vmatrix}$$

Now we have

$$\det \begin{vmatrix} -1 & 3 \\ 3 & -2 \end{vmatrix} = (-1)(-2) - (3)(3) = 2 - 9 = -7$$

$$\det \begin{vmatrix} 2 & 3 \\ 4 & -2 \end{vmatrix} = (2)(-2) - (3)(4) = -4 - 12 = -16$$

$$\det \begin{vmatrix} 2 & -1 \\ 4 & 3 \end{vmatrix} = (2)(3) - (4)(-1) = 6 + 4 = 10$$

and so

$$\det |A| = -7 + 32 - 10 = 15$$

Since this is nonzero we can apply Cramer's rule. We find

$$x = \frac{\det \begin{vmatrix} 4 & 2 & -1 \\ 3 & -1 & 3 \\ 5 & 3 & -2 \end{vmatrix}}{\det |A|}, \quad y = \frac{\det \begin{vmatrix} 1 & 4 & -1 \\ 2 & 3 & 3 \\ 4 & 5 & -2 \end{vmatrix}}{\det |A|}, \quad z = \frac{\det \begin{vmatrix} 1 & 2 & 4 \\ 2 & -1 & 3 \\ 4 & 3 & 5 \end{vmatrix}}{\det |A|}$$

Working out the first case explicitly, we find

$$\det \begin{vmatrix} 4 & 2 & -1 \\ 3 & -1 & 3 \\ 5 & 3 & -2 \end{vmatrix} = 4 \begin{vmatrix} -1 & 3 \\ 3 & -2 \end{vmatrix} - 2 \begin{vmatrix} 3 & 3 \\ 5 & -2 \end{vmatrix} - \begin{vmatrix} 3 & -1 \\ 5 & 3 \end{vmatrix}$$

$$= 4(2-9) - 2(-6-15) - (9+5)$$

$$= -28 + 42 - 14 = 0$$

For the other variables, using $\det |A| = 15$, we obtain (exercise)

$$y = \frac{45}{15} = 3$$

$$z = \frac{30}{15} = 2$$

Properties of Determinants

We now list some important properties of determinants:

- The determinant of a product of matrices is the product of their determinants, i.e., $\det |AB| = \det |A| \det |B|$.
- If the determinant of a matrix is nonzero, then that matrix has an inverse.
- If a matrix has a row or column of zeros, then $\det |A| = 0$.
- The determinant of a triangular matrix is the product of the diagonal elements.
- If the row or a column of a matrix B is multiplied by a scalar α to give a new matrix A, then $\det\det |A| = \alpha \det |B|$.

EXAMPLE 3-9
Show that $\det \det |AB| = \det |A| \det |B|$ for

$$A = \begin{pmatrix} a_{11} & a_{12} \\ a_{21} & a_{22} \end{pmatrix}, \quad B = \begin{pmatrix} b_{11} & b_{12} \\ b_{21} & b_{22} \end{pmatrix}$$

SOLUTION 3-9
We have

$$\det |A| = \det \left| \begin{pmatrix} a_{11} & a_{12} \\ a_{21} & a_{22} \end{pmatrix} \right| = a_{11}a_{22} - a_{21}a_{12}$$

$$\det |B| = \det \left| \begin{pmatrix} b_{11} & b_{12} \\ b_{21} & b_{22} \end{pmatrix} \right| = b_{11}b_{22} - b_{21}b_{12}$$

The product of these determinants is

$$\det |A| \det |B| = (a_{11}a_{22} - a_{21}a_{12})(b_{11}b_{22} - b_{21}b_{12})$$

$$= a_{11}a_{22}b_{11}b_{22} - a_{11}a_{22}b_{21}b_{12} - a_{21}a_{12}b_{11}b_{22} + a_{21}a_{12}b_{21}b_{12}$$

Now we compute the product of these matrices, and *then* the determinant. We have

$$AB = \begin{pmatrix} a_{11} & a_{12} \\ a_{21} & a_{22} \end{pmatrix} \begin{pmatrix} b_{11} & b_{12} \\ b_{21} & b_{22} \end{pmatrix} = \begin{pmatrix} a_{11}b_{11} + a_{12}b_{21} & a_{11}b_{12} + a_{12}b_{22} \\ a_{21}b_{11} + a_{22}b_{21} & a_{21}b_{12} + a_{22}b_{22} \end{pmatrix}$$

Therefore the determinant of the product is

$$\det |AB| = \det \begin{pmatrix} a_{11}b_{11} + a_{12}b_{21} & a_{11}b_{12} + a_{12}b_{22} \\ a_{21}b_{11} + a_{22}b_{21} & a_{21}b_{12} + a_{22}b_{22} \end{pmatrix}$$

$$= (a_{11}b_{11} + a_{12}b_{21})(a_{21}b_{12} + a_{22}b_{22})$$

$$- (a_{21}b_{11} + a_{22}b_{21})(a_{11}b_{12} + a_{12}b_{22})$$

Some simple algebra shows that this is

$$\det |AB| = a_{11}a_{22}b_{11}b_{22} - a_{11}a_{22}b_{21}b_{12} - a_{21}a_{12}b_{11}b_{22} + a_{21}a_{12}b_{21}b_{12}$$

$$= \det |A| \det |B|$$

EXAMPLE 3-10
Find the determinant of

$$B = \begin{bmatrix} 1 & -2 & 4 \\ 0 & 6 & -2 \\ 0 & 0 & 1 \end{bmatrix}$$

in two ways.

SOLUTION 3-10

First we compute the determinant using the brute force method:

$$\det |B| = \det \begin{bmatrix} 1 & -2 & 4 \\ 0 & 6 & -2 \\ 0 & 0 & 1 \end{bmatrix} = \begin{vmatrix} 6 & -2 \\ 0 & 1 \end{vmatrix} + 2 \begin{vmatrix} 0 & -2 \\ 0 & 1 \end{vmatrix} + 4 \begin{vmatrix} 0 & 6 \\ 0 & 0 \end{vmatrix}$$

The properties of determinants tell us that if a row or column of a matrix is zero, then the determinant is zero. Therefore the second and third determinants are zero, leaving

$$\det |B| = \begin{vmatrix} 6 & -2 \\ 0 & 1 \end{vmatrix} = (6)(1) - (0)(-2) = 6$$

To compute the determinant a second way, we note that the matrix is triangular and compute the determinant by multiplying the diagonal elements together:

$$\det |B| = (1)(6)(1) = 6$$

EXAMPLE 3-11

Prove that if the first row of a third-order square matrix is all zeros, the determinant is zero.

SOLUTION 3-11

The proof is straightforward. We write the matrix as

$$A = \begin{bmatrix} 0 & 0 & 0 \\ a & b & c \\ d & e & f \end{bmatrix}$$

and we see immediately that

$$\det |A| = \begin{vmatrix} 0 & 0 & 0 \\ a & b & c \\ d & e & f \end{vmatrix} = (0) \begin{vmatrix} b & c \\ e & f \end{vmatrix} - (0) \begin{vmatrix} a & c \\ d & f \end{vmatrix} + (0) \begin{vmatrix} a & b \\ d & e \end{vmatrix} = 0$$

By showing this for the other two rows, we can demonstrate that this result is true in general for third-order matrices.

EXAMPLE 3-12

Prove that if the first column of a matrix B is multiplied by a scalar α to give a new matrix A, then $\det |A| = \alpha \det |B|$ for a third-order matrix.

SOLUTION 3-12
We take

$$B = \begin{bmatrix} a_1 & a_2 & a_3 \\ b_1 & b_2 & b_3 \\ c_1 & c_2 & c_3 \end{bmatrix}$$

Using the formula for the determinant of a third-order matrix, we have

$$\det |B| = a_1 \det \begin{vmatrix} b_2 & b_3 \\ c_2 & c_3 \end{vmatrix} - a_2 \det \begin{vmatrix} b_1 & b_3 \\ c_1 & c_3 \end{vmatrix} + a_3 \det \begin{vmatrix} b_1 & b_2 \\ c_1 & c_2 \end{vmatrix}$$

$$= a_1 (b_2 c_3 - c_2 b_3) - a_2 (b_1 c_3 - c_1 b_3) + a_3 (b_1 c_2 - c_1 b_2)$$

Now

$$A = \begin{bmatrix} \alpha a_1 & a_2 & a_3 \\ \alpha b_1 & b_2 & b_3 \\ \alpha c_1 & c_2 & c_3 \end{bmatrix}$$

and so

$$\det |A| = \alpha a_1 \det \begin{vmatrix} b_2 & b_3 \\ c_2 & c_3 \end{vmatrix} - a_2 \det \begin{vmatrix} \alpha b_1 & b_3 \\ \alpha c_1 & c_3 \end{vmatrix} + a_3 \det \begin{vmatrix} \alpha b_1 & b_2 \\ \alpha c_1 & c_2 \end{vmatrix}$$

$$= \alpha a_1 (b_2 c_3 - c_2 b_3) - a_2 (\alpha b_1 c_3 - \alpha c_1 b_3) + a_3 (\alpha b_1 c_2 - \alpha c_1 b_2)$$

$$= \alpha [a_1 (b_2 c_3 - c_2 b_3) - a_2 (b_1 c_3 - c_1 b_3) + a_3 (b_1 c_2 - c_1 b_2)]$$

$$= \alpha \det |B|$$

Finding the Inverse of a Matrix

If the determinant of a matrix is nonzero, the inverse exists. We calculate the inverse of A by using the determinant and the *adjugate*, a matrix whose (i, j) entry is found by calculating the *cofactor* of the entry (i, j) of A.

THE MINOR

Let A_{mn} be the submatrix formed from A by eliminating the mth row and nth column of A. The minor associated with entry (m, n) of A is the determinant of A_{mn}.

EXAMPLE 3-13
Let

$$A = \begin{bmatrix} -2 & 3 & 1 \\ 0 & 4 & 5 \\ 2 & 1 & 4 \end{bmatrix}$$

Find the minors for (1, 1) and (2, 3).

SOLUTION 3-13
To find the minor for (1, 1), we eliminate the first row and the first column of the matrix to give the submatrix

$$A_{11} = \begin{bmatrix} 4 & 5 \\ 1 & 4 \end{bmatrix}$$

The minor associated with (1, 1) is the determinant of this matrix:

$$\det |A_{11}| = \det \begin{vmatrix} 4 & 5 \\ 1 & 4 \end{vmatrix} = (4)(4) - (1)(5) = 16 - 5 = 11$$

Now to find the minor for (2, 3), we cross out row 2 and column 3 of the matrix A to create the submatrix

$$A_{23} = \begin{bmatrix} -2 & 3 \\ 2 & 1 \end{bmatrix}$$

The minor is the determinant of this matrix:

$$\det \begin{vmatrix} -2 & 3 \\ 2 & 1 \end{vmatrix} = -2 - 6 = -8$$

THE COFACTOR

To find the *cofactor* for entry (m, n) of a matrix A, we calculate the signed minor for entry (m, n), which is given by

$$(-1)^{m+n} \det |A_{mn}|$$

We denote the cofactor by a_{mn}.

EXAMPLE 3-14
Find the cofactors corresponding to the minors in the previous example.

SOLUTION 3-14
We have

$$a_{11} = (-1)^{1+1} \det|A_{11}| = (-1)^2 (11) = 11$$

and

$$a_{23} = (-1)^{2+3} \det|A_{23}| = (-1)^5 (-8) = (-1)(-8) = 8$$

THE ADJUGATE OF A MATRIX

The adjugate of a matrix A is the matrix of the cofactors. For a 3×3 matrix

$$\text{adj}(A) = \begin{bmatrix} a_{11} & a_{21} & a_{31} \\ a_{12} & a_{22} & a_{32} \\ a_{13} & a_{23} & a_{33} \end{bmatrix}$$

Notice that the row and column indices are reversed. So we calculate the cofactor for the (i, j) entry of matrix A and then we use this for the (j, i) entry of the adjugate.

THE INVERSE

If the determinant of A is nonzero, then

$$A^{-1} = \frac{1}{\det|A|} \text{adj}(A)$$

EXAMPLE 3-15
Find the inverse of the matrix

$$A = \begin{bmatrix} 2 & 4 \\ 6 & 8 \end{bmatrix}$$

SOLUTION 3-15
First we calculate the determinant

$$\det|A| = \det \begin{vmatrix} 2 & 4 \\ 6 & 8 \end{vmatrix} = (2)(8) - (6)(4) = 16 - 24 = -8$$

The determinant is nonzero; therefore, the inverse exists.

For a 2×2 matrix, submatrices used to calculate the minors are just numbers. To calculate A_{11}, we eliminate the first row and first column of the matrix:

$$A_{11} = \begin{pmatrix} a_{11} & a_{12} \\ a_{21} & a_{22} \end{pmatrix} = 8$$

To calculate A_{12}, we eliminate the first row and second column:

$$A_{12} = \begin{pmatrix} a_{11} & a_{12} \\ a_{21} & a_{22} \end{pmatrix} = 6$$

To calculate A_{21}, we eliminate the second row and first column:

$$A_{21} = \begin{pmatrix} a_{11} & a_{12} \\ a_{21} & a_{22} \end{pmatrix} = 4$$

Finally, to find A_{22}, we eliminate the second row and second column:

$$A_{22} = \begin{pmatrix} a_{11} & a_{12} \\ a_{21} & a_{22} \end{pmatrix} = 2$$

We calculate each of the cofactors, noting that the determinant of a number is just that number. In other words, $\det(\alpha) = \alpha$. And so we have

$$a_{11} = (-1)^{1+1} \det|A_{11}| = 8$$
$$a_{12} = (-1)^{1+2} \det|A_{12}| = -6$$
$$a_{21} = (-1)^{2+1} \det|A_{21}| = -4$$
$$a_{22} = (-1)^{2+2} \det|A_{22}| = 2$$
$$\Rightarrow adj(A) = \begin{bmatrix} a_{11} & a_{21} \\ a_{12} & a_{22} \end{bmatrix} = \begin{bmatrix} 8 & -4 \\ -6 & 2 \end{bmatrix}$$

Dividing through by the determinant, we obtain

$$A^{-1} = \frac{1}{\det|A|} adj(A) = -\frac{1}{8}\begin{bmatrix} 8 & -4 \\ -6 & 2 \end{bmatrix} = \begin{bmatrix} -1 & \dfrac{1}{2} \\ \dfrac{3}{4} & -\dfrac{1}{4} \end{bmatrix}$$

We verify that this is in fact the inverse:

$$A = \begin{bmatrix} 2 & 4 \\ 6 & 8 \end{bmatrix} \begin{bmatrix} -1 & \frac{1}{2} \\ \frac{3}{4} & -\frac{1}{4} \end{bmatrix} = \begin{bmatrix} (2)(-1) + (4)\left(\frac{3}{4}\right) & (2)\left(\frac{1}{2}\right) + (4)\left(-\frac{1}{4}\right) \\ (6)(-1) + (8)\left(\frac{3}{4}\right) & (6)\left(\frac{1}{2}\right) + (8)\left(-\frac{1}{4}\right) \end{bmatrix}$$

$$= \begin{bmatrix} -2+3 & 1-1 \\ -6+6 & 3-2 \end{bmatrix} = \begin{bmatrix} 1 & 0 \\ 0 & 1 \end{bmatrix} = I$$

Quiz

1. Find the determinant of the matrix

$$A = \begin{bmatrix} 1 & 9 \\ 2 & 5 \end{bmatrix}$$

2. Find the determinant of the matrix

$$B = \begin{bmatrix} -7 & 0 & 4 \\ 2 & 1 & 9 \\ 6 & 5 & 1 \end{bmatrix}$$

3. Show that $\det(AB) = \det(A)\det(B)$ for

$$A = \begin{bmatrix} 8 & -1 \\ 3 & -6 \end{bmatrix}, \qquad B = \begin{bmatrix} -4 & 1 \\ 2 & 6 \end{bmatrix}$$

4. If possible, solve the linear system

$$x + 2y = 7$$
$$3x - 4y = 9$$

using Cramer's rule.

5. If possible, find the point of intersection for the planes

$$7x - 2y + z = 15$$
$$x + y - 3z = 4$$
$$2x - y + 5z = 2$$

6. Let

$$A = \begin{bmatrix} -2 & 1 & 0 \\ 2 & 6 & 2 \\ 1 & 8 & 4 \end{bmatrix}$$

Find the cofactors for this matrix.

7. Find the adjugate of the matrix A in the previous problem.

8. For the matrix of the problem 6, calculate the determinant and find out if the inverse exists. If so, find the inverse.

9. Consider an arbitrary 2×2 matrix

$$\begin{pmatrix} a_{11} & a_{12} \\ a_{21} & a_{22} \end{pmatrix}$$

Write down the transpose of this matrix and see if you can determine a relationship between the determinant of the original matrix and the determinant of its transpose.

10. Find the determinants of

$$A = \begin{bmatrix} 2 & 1 & -1 \\ -1 & 4 & 4 \\ 5 & 1 & -1 \end{bmatrix}, \qquad B = \begin{bmatrix} 2 & -1 & 1 \\ -1 & 4 & 4 \\ 5 & -1 & 1 \end{bmatrix}$$

Is $\det |A| = -\det |B|$?

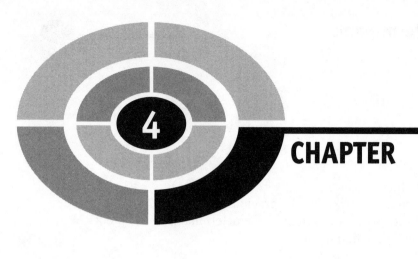

CHAPTER

Vectors

The reader is probably familiar with vectors from their use in physics and engineering. A vector is a quantity that has both *magnitude* and *direction*. Mathematically, we can represent a vector graphically in the plane by a directed line segment or arrow that has its tail at one point and the head of the arrow at a second point, as illustrated in Fig. 4-1.

Two vectors can be added together using geometric means by using the *parallelogram law*. To add two vectors **u** and **v**, we place the tail of **v** at the head of **u** and then draw a line from the tail of **u** to the tip of **v**. This new vector is **u** + **v** (see Fig. 4-2).

While we can work with vectors geometrically, we aren't going to spend anymore time thinking about such notions because in linear algebra we work with abstract vectors that encapsulate the fundamental properties of the geometric vectors we are familiar with from physics. Let's think about what those properties are. We can

- Add or subtract two vectors, giving us another vector.

Fig. 4-1. An example of a vector.

- Multiply a vector by a scalar, giving another vector that has been stretched or shrunk.
- Form a scalar product or number from two vectors by computing their dot or inner product.
- Find the angle between two vectors by computing their dot product.
- Describe a zero vector, which, when added to another vector, leaves that vector alone.
- Find an inverse of any vector, which, when added to that vector, gives the zero vector.
- Represent a vector with respect to a set of basis vectors, which means we can represent the vector by a set of numbers.

This last property is going to be of fundamental importance in linear algebra, where we will work with abstract vectors by manipulating their components. Let's refresh our memory a bit by considering two basis vectors in the plane, one points in the x direction and the other points in the y direction (see Fig. 4-3). Remember the basis vectors have *unit length*.

A given vector in the $x-y$ plane can be decomposed into components along the x and y axes (see Fig. 4-4).

The way we write this mathematically is that we *expand* the vector **u** with respect to the basis $\{\hat{x}, \hat{y}, \hat{z}\}$. The values u_x, u_y are the components of the vector. The expansion of this vector is

$$\vec{u} = u_x \hat{x} + u_y \hat{y}$$

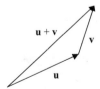

Fig. 4-2. Addition of two vectors.

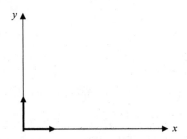

Fig. 4-3. Basis vectors in the $x-y$ plane.

Vectors are added together by adding components

$$\vec{u} = u_x\hat{x} + u_y\hat{y} + u_z\hat{z}$$
$$\vec{v} = v_x\hat{x} + v_y\hat{y} + v_z\hat{z}$$
$$\Rightarrow \vec{u} + \vec{v} = (u_x + v_x)\hat{x} + \left(u_y + v_y\right)\hat{y} + (u_z + v_z)\hat{z}$$

The dot product between two vectors is found to be

$$\vec{u} \cdot \vec{v} = u_x v_x + u_y v_y + u_z v_z$$

and the length of a vector is found by taking the dot product of a vector with itself, i.e.,

$$\|\vec{u}\| = \sqrt{u_x^2 + u_y^2 + u_z^2}$$

Now let's see how we can abstract notions like these to work with vectors in complex vector space or n-dimensional spaces.

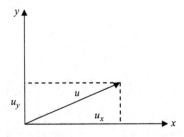

Fig. 4-4. Decomposing some vector **u** into x and y components.

Vectors in $\mathbb{R}^n$

Consider the set of *n-tuples* of real numbers, which are nothing more than lists of n numbers. For example

$$u = (u_1, u_2, \ldots, u_n)$$

is a valid *n*-tuple. The numbers u_i are called the *components* of u. A specific example is

$$u = (3, -2, 4)$$

where μ is a vector in $\mathbb{R}^3$. We now consider some basic operations on vectors in $\mathbb{R}^n$.

Vector Addition

We can also represent vectors by a list of numbers arranged in a column. This is called a *column vector*. For example, we write a vector u in $\mathbb{R}^n$ as

$$u = \begin{bmatrix} u_1 \\ u_2 \\ \vdots \\ u_n \end{bmatrix}$$

Vector addition is carried out componentwise. Specifically, given two vectors that belong to the vector space $\mathbb{R}^n$

$$u = \begin{bmatrix} u_1 \\ u_2 \\ \vdots \\ u_n \end{bmatrix}, \qquad v = \begin{bmatrix} v_1 \\ v_2 \\ \vdots \\ v_n \end{bmatrix}$$

We form the sum $u + v$ by adding component by component in the following way:

$$u + v = \begin{bmatrix} u_1 \\ u_2 \\ \vdots \\ u_n \end{bmatrix} + \begin{bmatrix} v_1 \\ v_2 \\ \vdots \\ v_n \end{bmatrix} = \begin{bmatrix} u_1 + v_1 \\ u_2 + v_2 \\ \vdots \\ u_n + v_n \end{bmatrix}$$

Notice that if u and v are valid lists of real numbers, then so is their sum. Therefore the sum $u + v$ is also a vector in $\mathbb{R}^n$. We say that $\mathbb{R}^n$ is closed under vector addition.

EXAMPLE 4-1
Consider two vectors that belong to $\mathbb{R}^3$

$$u = \begin{bmatrix} -1 \\ 2 \\ 5 \end{bmatrix}, \qquad v = \begin{bmatrix} 7 \\ 8 \\ -1 \end{bmatrix}$$

Compute the vector formed by the sum $u + v$.

SOLUTION 4-1
We form the sum by adding components:

$$u + v = \begin{bmatrix} -1 \\ 2 \\ 5 \end{bmatrix} + \begin{bmatrix} 7 \\ 8 \\ -1 \end{bmatrix} = \begin{bmatrix} -1 + 7 \\ 2 + 8 \\ 5 + (-1) \end{bmatrix} = \begin{bmatrix} 6 \\ 10 \\ 4 \end{bmatrix}$$

We can also consider complex vector spaces. A vector in $\mathbb{C}^n$ is also an n-tuple, but this time we allow the elements or components of the vector to be complex numbers. Therefore two vectors in $\mathbb{C}^3$ are

$$v = \begin{bmatrix} 2 + i \\ 3 \\ 4i \end{bmatrix}, \qquad w = \begin{bmatrix} 0 \\ 1 \\ -i \end{bmatrix}$$

Most operations on vectors that belong to a complex vector space are carried out in essentially the same way. For example, we can add together these vectors

component by component:

$$v + w = \begin{bmatrix} 2+i \\ 3 \\ 4i \end{bmatrix} + \begin{bmatrix} 0 \\ 1 \\ -i \end{bmatrix} = \begin{bmatrix} 2+i \\ 4 \\ 3i \end{bmatrix}$$

Scalar Multiplication

Let u be a vector in some vector space

$$u = \begin{bmatrix} u_1 \\ u_2 \\ \vdots \\ u_n \end{bmatrix}$$

and α be a scalar. The scalar product of α and u is given by

$$\alpha u = \alpha \begin{bmatrix} u_1 \\ u_2 \\ \vdots \\ u_n \end{bmatrix} = \begin{bmatrix} \alpha u_1 \\ \alpha u_2 \\ \vdots \\ \alpha u_n \end{bmatrix}$$

If we are dealing with a real vector space, then the scalar α must be a real number. For a complex vector space, α can be real or complex.

EXAMPLE 4-2

Let $\alpha = 3$ and suppose that

$$u = \begin{bmatrix} -1 \\ 4 \\ 5 \end{bmatrix}, \qquad v = \begin{bmatrix} 0 \\ 2 \\ 5 \end{bmatrix}$$

Find $\alpha u - v$.

SOLUTION 4-2

The scalar product is

$$\alpha u = (3) \begin{bmatrix} -1 \\ 4 \\ 5 \end{bmatrix} = \begin{bmatrix} (3)(-1) \\ (3)(4) \\ (3)(5) \end{bmatrix} = \begin{bmatrix} -3 \\ 12 \\ 15 \end{bmatrix}$$

Therefore

$$\alpha u - v = \begin{bmatrix} -3 \\ 12 \\ 15 \end{bmatrix} - \begin{bmatrix} 0 \\ 2 \\ 5 \end{bmatrix} = \begin{bmatrix} -3-0 \\ 12-2 \\ 15-5 \end{bmatrix} = \begin{bmatrix} -3 \\ 10 \\ 10 \end{bmatrix}$$

EXAMPLE 4-3
Let

$$u = \begin{bmatrix} 2i \\ 6 \end{bmatrix}$$

and $\alpha = 3 + 2i$. Find αu.

SOLUTION 4-3
Scalar multiplication in a complex vector space also proceeds component by component. Therefore we have

$$\alpha u = (3+2i) \begin{bmatrix} 2i \\ 6 \end{bmatrix} = \begin{bmatrix} (3)(2i) + (2i)(2i) \\ (3+2i)(6) \end{bmatrix} = \begin{bmatrix} -4+6i \\ 18+12i \end{bmatrix}$$

If you're rusty with complex numbers recall that $i^2 = -1$, $(i)(-i) = +1$.
   Subtraction of two vectors proceeds in an analogous way, as the next example shows.

EXAMPLE 4-4
In the vector space $\mathbb{R}^4$ find $u - v$ for

$$u = \begin{bmatrix} 2 \\ -1 \\ 3 \\ 4 \end{bmatrix}, \qquad v = \begin{bmatrix} 1 \\ 2 \\ -5 \\ 6 \end{bmatrix}$$

SOLUTION 4-4
Working component by component, we obtain

$$u - v = \begin{bmatrix} 2 \\ -1 \\ 3 \\ 4 \end{bmatrix} - \begin{bmatrix} 1 \\ 2 \\ -5 \\ 6 \end{bmatrix} = \begin{bmatrix} 2-1 \\ -1-2 \\ 3+5 \\ 4-6 \end{bmatrix} = \begin{bmatrix} 1 \\ -3 \\ 8 \\ -2 \end{bmatrix}$$

The Zero Vector

To represent the zero vector in an abstract vector space, we basically have just a list of zeros. The property that a zero vector must satisfy is

$$u + 0 = 0 + u = u$$

for any vector that belongs to a given vector space. So, we have

$$u + 0 = \begin{bmatrix} u_1 \\ u_2 \\ \vdots \\ u_n \end{bmatrix} + \begin{bmatrix} 0 \\ 0 \\ \vdots \\ 0 \end{bmatrix} = \begin{bmatrix} u_1 + 0 \\ u_2 + 0 \\ \vdots \\ u_n + 0 \end{bmatrix} = \begin{bmatrix} u_1 \\ u_2 \\ \vdots \\ u_n \end{bmatrix} = u$$

The inverse of a vector is found by negating all the components. If

$$u = \begin{bmatrix} u_1 \\ u_2 \\ \vdots \\ u_n \end{bmatrix}$$

then

$$-u = \begin{bmatrix} -u_1 \\ -u_2 \\ \vdots \\ -u_n \end{bmatrix}$$

We see immediately that

$$u + (-u) = \begin{bmatrix} u_1 \\ u_2 \\ \vdots \\ u_n \end{bmatrix} + \begin{bmatrix} -u_1 \\ -u_2 \\ \vdots \\ -u_n \end{bmatrix} = \begin{bmatrix} u_1 - u_1 \\ u_2 - u_2 \\ \vdots \\ u_n - u_n \end{bmatrix} = \begin{bmatrix} 0 \\ 0 \\ \vdots \\ 0 \end{bmatrix} = 0$$

The Transpose of a Vector

We now consider the *transpose* of a vector in $\mathbb{R}^n$, which is a row vector. For a vector

$$u = \begin{bmatrix} u_1 \\ u_2 \\ \vdots \\ u_n \end{bmatrix}$$

the transpose is denoted by

$$u^T = \begin{bmatrix} u_1 \, u_2 \cdots u_n \end{bmatrix}$$

EXAMPLE 4-5
Find the transpose of the vector

$$v = \begin{bmatrix} 1 \\ -2 \\ 5 \end{bmatrix}$$

SOLUTION 4-5
The transpose is found by writing the components of the vector in a row. Therefore we have

$$v^T = \begin{bmatrix} 1 & -2 & 5 \end{bmatrix}$$

EXAMPLE 4-6
Find the transpose of

$$u = \begin{bmatrix} 1 \\ 0 \\ -2 \\ 1 \end{bmatrix}$$

SOLUTION 4-6
We write the components of u as a row vector

$$u^T = \begin{bmatrix} 1 & 0 & -2 & 1 \end{bmatrix}$$

For vectors in a complex vector space, the situation is slightly more complicated. We call the equivalent vector the *conjugate* and we need to apply two steps to calculate it:

- Take the transpose of the vector.
- Compute the complex conjugate of each component.

The conjugate of a vector in a complex vector space is written as $u^\dagger$. If you don't recall complex numbers, the complex conjugate is found by letting $i \to -i$. In this book we denote the complex conjugate operation by *. Therefore the complex conjugate of α is written as α^*. The best way to learn what to do is to look at a couple of examples.

EXAMPLE 4-7
Let

$$u = \begin{bmatrix} 2i \\ 5 \end{bmatrix}$$

Calculate $u^\dagger$.

SOLUTION 4-7
First we take the transpose of the vector

$$u^T = \begin{bmatrix} 2i & 5 \end{bmatrix}$$

Now we take the complex conjugate, letting $i \to -i$:

$$u^\dagger = \begin{bmatrix} 2i & 5 \end{bmatrix}^* = \begin{bmatrix} -2i & 5 \end{bmatrix}$$

EXAMPLE 4-8
Find the conjugate of

$$v = \begin{bmatrix} 2+i \\ 3i \\ 4-5i \end{bmatrix}$$

SOLUTION 4-8
We take the transpose to write v as a row vector

$$v^T = \begin{bmatrix} 2+i & 3i & 4-5i \end{bmatrix}$$

Now take the complex conjugate of each component to obtain

$$v^\dagger = \begin{bmatrix} 2+i & 3i & 4-5i \end{bmatrix}^* = \begin{bmatrix} 2-i & -3i & 4+5i \end{bmatrix}$$

The Dot or Inner Product

We needed to learn how to write column vectors as row vectors for real and complex vector spaces because this makes computing inner products much easier. The inner product is a number and so it is also known as the *scalar product*. In a real vector space, the scalar product between two vectors

$$u = \begin{bmatrix} u_1 \\ u_2 \\ \vdots \\ u_n \end{bmatrix}, \qquad v = \begin{bmatrix} v_1 \\ v_2 \\ \vdots \\ v_n \end{bmatrix}$$

is computed in the following way:

$$(u, v) = \begin{bmatrix} u_1 & u_2 & \cdots & u_n \end{bmatrix} \begin{bmatrix} v_1 \\ v_2 \\ \vdots \\ v_n \end{bmatrix} = u_1 v_1 + u_2 v_2 + \cdots + u_n v_n = \sum_{i=1}^{n} u_i v_i$$

EXAMPLE 4-9
Let

$$u = \begin{bmatrix} 2 \\ -1 \\ 3 \end{bmatrix}, \qquad v = \begin{bmatrix} 4 \\ 5 \\ -6 \end{bmatrix}$$

and compute their dot product.

SOLUTION 4-9
We have

$$(u, v) = \begin{bmatrix} 2 & -1 & 3 \end{bmatrix} \begin{bmatrix} 4 \\ 5 \\ -6 \end{bmatrix} = (2)(4) + (-1)(5) + (3)(-6)$$

$$= 8 - 5 - 18 = -15$$

If the inner product of two vectors is zero, we say that the vectors are *orthogonal*.

EXAMPLE 4-10
Show that

$$u = \begin{bmatrix} 1 \\ -2 \\ 2 \end{bmatrix}, \qquad v = \begin{bmatrix} 2 \\ 5 \\ 4 \end{bmatrix}$$

are orthogonal.

SOLUTION 4-10
The inner product is

$$(u, v) = \begin{bmatrix} 1 & -2 & 2 \end{bmatrix} \begin{bmatrix} 2 \\ 5 \\ 4 \end{bmatrix} = (1)(2) + (-2)(5) + (2)(4) = 2 - 10 + 8 = 0$$

To compute the inner product in a complex vector space, we compute the conjugate of the first vector. We use the notation

$$(u, v) = \begin{bmatrix} u_1^* & u_2^* & \cdots & u_n^* \end{bmatrix} \begin{bmatrix} v_1 \\ v_2 \\ \vdots \\ v_n \end{bmatrix} = u_1^* v_1 + u_2^* v_2 + \cdots + u_n^* v_n = \sum_{i=1}^{n} u_i^* v_i$$

EXAMPLE 4-11
Find the inner product of

$$u = \begin{bmatrix} 2i \\ 6 \end{bmatrix}, \qquad v = \begin{bmatrix} 3 \\ 5i \end{bmatrix}$$

SOLUTION 4-11
Taking the conjugate of u, we obtain

$$u^{\dagger} = \begin{bmatrix} -2i & 6 \end{bmatrix}$$

Therefore we have

$$(u, v) = \begin{bmatrix} -2i & 6 \end{bmatrix} \begin{bmatrix} 3 \\ 5i \end{bmatrix} = (-2i)(3) + (6)(5i) = -6i + 30i = 24i$$

Note that the inner product is a linear operation, and so

$$(u + v, w) = (u, w) + (v, w)$$
$$(u, v + w) = (u, v) + (u, w)$$
$$(\alpha u, v) = \alpha (u, v)$$
$$(u, v) = (v, u)$$

In a complex vector space, we have

$$(\alpha u, v) = \alpha^* (u, v)$$
$$(u, \beta v) = \beta (u, v)$$
$$(u, v) = (v, u)^*$$

The Norm of a Vector

We carry over the notion of length to abstract vector spaces through the *norm*. The norm is written as $\|u\|$ and is defined as the nonnegative square root of the dot product (u, u). More specifically, we have

$$\|u\| = \sqrt{(u, u)} = \sqrt{u_1^2 + u_2^2 + \cdots + u_n^2}$$

The norm must be a real number to have any meaning as a length. This is why we compute the conjugate of the vector in the first slot of the inner product for a complex vector space. We illustrate this more clearly with an example.

EXAMPLE 4-12
Find the norm of

$$v = \begin{bmatrix} 2i \\ 4 \\ 1 + i \end{bmatrix}$$

SOLUTION 4-12

We first compute the conjugate

$$v^\dagger = \begin{bmatrix} 2i \\ 4 \\ 1+i \end{bmatrix}^\dagger = \begin{bmatrix} -2i & 4 & 1-i \end{bmatrix}$$

The inner product is

$$(v, v) = \begin{bmatrix} -2i & 4 & 1-i \end{bmatrix} \begin{bmatrix} 2i \\ 4 \\ 1+i \end{bmatrix} = (-2i)(2i) + (4)(4) + (1-i)(1+i)$$

$$= 4 + 16 + 2 = 22$$

The norm of the vector is the positive square root of this quantity:

$$\|v\| = \sqrt{(v, v)} = \sqrt{22}$$

Note that

$$(u, u) \geq 0$$

For any vector u, with equality only for the zero vector.

Unit Vectors

A unit vector is a vector that has a norm that is equal to 1. We can construct a unit vector from any vector v by writing

$$\tilde{v} = \frac{v}{\|v\|}$$

EXAMPLE 4-13

A vector in a real vector space is

$$w = \begin{bmatrix} 2 \\ -1 \end{bmatrix}$$

Use this vector to construct a unit vector.

SOLUTION 4-13

The inner product is

$$(w, w) = (2)(2) + (-1)(-1) = 4 + 1 = 5$$

The norm of this vector is the positive square root:

$$\|w\| = \sqrt{(w, w)} = \sqrt{5}$$

We can construct a unit vector by dividing w by its norm:

$$u = \frac{w}{\|w\|} = \frac{1}{\sqrt{5}} w = \frac{1}{\sqrt{5}} \begin{bmatrix} 2 \\ -1 \end{bmatrix} = \begin{bmatrix} \frac{2}{\sqrt{5}} \\ -\frac{1}{\sqrt{5}} \end{bmatrix}$$

We call this procedure *normalization* or say we are *normalizing* the vector.

The Angle between Two Vectors

The angle between two vectors u and v is

$$\cos \theta = \frac{(u, v)}{\|u\| \, \|v\|}$$

Two Theorems Involving Vectors

The Cauchy–Schwartz inequality states that

$$|(u, v)| \leq \|u\| \, \|v\|$$

and the triangle inequality says

$$\|u + v\| \leq \|u\| + \|v\|$$

Distance between Two Vectors

We can carry over a notion of "distance" between two vectors. This is given by

$$d(u, v) = \|u - v\| = \sqrt{(u_1 - v_1)^2 + (u_2 - v_2)^2 + \cdots + (u_n - v_n)^2}$$

EXAMPLE 4-14
Find the distance between

$$u = \begin{bmatrix} 2 \\ -1 \\ 2 \end{bmatrix}, \qquad v = \begin{bmatrix} 1 \\ 3 \\ 4 \end{bmatrix}$$

SOLUTION 4-14
The difference between the vectors is

$$u - v = \begin{bmatrix} 2 \\ -1 \\ 2 \end{bmatrix} - \begin{bmatrix} 1 \\ 3 \\ 4 \end{bmatrix} = \begin{bmatrix} 2 - 1 \\ -1 - 3 \\ 2 - 4 \end{bmatrix} = \begin{bmatrix} 1 \\ -4 \\ -2 \end{bmatrix}$$

The inner product is

$$(u - v, u - v) = (1)^2 + (-4)^2 + (-2)^2 = 1 + 16 + 4 = 21$$

and so the distance function gives

$$d(u, v) = \|u - v\| = \sqrt{(u - v, u - v)} = \sqrt{21}$$

Quiz

1. Construct the sum and difference of the vectors

$$v = \begin{bmatrix} -2 \\ 4 \end{bmatrix}, \qquad w = \begin{bmatrix} 1 \\ 8 \end{bmatrix}$$

2.   Find the scalar multiplication of the vector

$$u = \begin{bmatrix} 2 \\ -1 \\ 4 \end{bmatrix}$$

by $k = 3$.

3.   Using the rules for vector addition and scalar multiplication, write the vector

$$a = \begin{bmatrix} 2 \\ -3 \\ 4 \end{bmatrix}$$

in terms of the vectors

$$e_1 = \begin{bmatrix} 1 \\ 0 \\ 0 \end{bmatrix}, \qquad e_2 = \begin{bmatrix} 0 \\ 1 \\ 0 \end{bmatrix}, \qquad e_3 = \begin{bmatrix} 0 \\ 0 \\ 1 \end{bmatrix}$$

The vectors e_i are called the *standard basis* of $\mathbb{R}^3$.

4.   Find the inner product of

$$u = \begin{bmatrix} 2 \\ 4i \end{bmatrix}, \qquad v = \begin{bmatrix} -1 \\ 3 \end{bmatrix}$$

5.   Find the norm of the vectors

$$a = \begin{bmatrix} 2 \\ -2 \end{bmatrix}, \qquad b = \begin{bmatrix} 1 \\ -i \\ 2 \end{bmatrix}, \qquad c = \begin{bmatrix} 8i \\ 2 \\ i \end{bmatrix}$$

6.   Normalize the vectors

$$a = \begin{bmatrix} 2 \\ 3 \\ -1 \end{bmatrix}, \qquad u = \begin{bmatrix} 1+i \\ 4-i \end{bmatrix}$$

7. Let

$$u = \begin{bmatrix} 2 \\ -1 \end{bmatrix}, \qquad v = \begin{bmatrix} 4 \\ 5 \end{bmatrix}, \qquad w = \begin{bmatrix} -1 \\ 1 \end{bmatrix}$$

Find
(a) $u + 2v - w$
(b) $3w$
(c) $-2u + 5v + 7w$
(d) The norm of each vector
(e) Normalize each vector

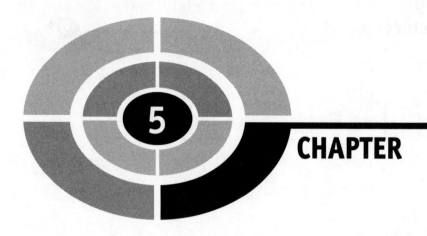

5 CHAPTER

Vector Spaces

A *vector space V* is a set of elements $u, v, w, \ldots$ called *vectors* that satisfy the following axioms:

- A vector space is closed under addition. This means there exists an operation called addition such that the sum of two vectors, given by $w = u + v$ is another vector that belongs to V.
- A vector space is closed under scalar multiplication. If $u \in V$ then so is αu, where α is a number.
- Vector addition is associative, meaning that
  $u + (v + w) = (u + v) + w$.
- Vector addition is commutative, i.e., $u + v = v + u$.
- There exists a unique *zero vector* that satisfies $0 + u = u + 0 = u$.
- There exists an additive inverse such that $u + (-u) = (-u) + u = 0$.
- Scalar multiplication is distributive, i.e., $\alpha (u + v) = \alpha u + \alpha v$.
- Scalar multiplication is associative, meaning $\alpha (\beta u) = (\alpha \beta) u$.
- There exists an identity element I such that $Iu = uI = u$ for each $u \in V$.

These are general mathematical properties that apply to a wide range of objects, not just geometric vectors. Certain types of functions can form a vector space, for example. Often one is asked to determine whether a given collection of elements is a vector space.

EXAMPLE 5-1

Does the function

$$4x - y = 7$$

constitute a vector space?

SOLUTION 5-1

This function is the line shown in Fig. 5-1.

We can show that this line is not a vector space by showing that it does not satisfy closure under addition. Let x_1, x_2 be two points on the x axis and y_1, y_2 be two points on the y axis such that

$$4x_1 - y_1 = 7$$
$$4x_2 - y_2 = 7$$

Adding the two elements, we obtain

$$4(x_1 + x_2) - (y_1 + y_2) = 14 \neq 7$$

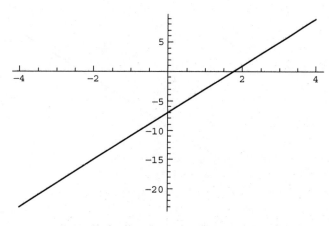

Fig. 5-1. Is the line $4x - y = 7$ a vector space?

For closure to be satisfied, $4(x_1 + x_2) - (y_1 + y_2)$ must also sum to 7, which it does not. Therefore the line $4x - y = 7$ is *not* a vector space. We can also see this by noting it is not closed under scalar multiplication. Again suppose we had a point (x_1, y_1) such that

$$4x_1 - y_1 = 7$$

This means that

$$3(4x_1 - y_1) = 12x_1 - 3y_1$$

But on the right-hand side, we have $3 \times 7 = 21$ and so the result is not 7.

EXAMPLE 5-2
Show that the line

$$x - 2y = 0$$

is closed under addition and scalar multiplication.

SOLUTION 5-2
This is the line through the origin shown in Fig. 5-2.
    We suppose that (x_1, y_1) and (x_2, y_2) are two points such that

$$x_1 - 2y_1 = 0$$
$$x_2 - 2y_2 = 0$$

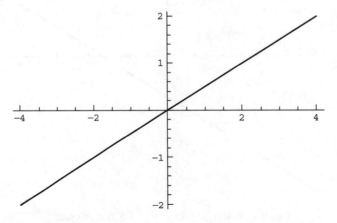

Fig. 5-2. The line $x - 2y = 0$ does constitute a vector space.

Adding we find

$$(x_1 + x_2) - 2(y_1 + y_2) = 0$$

Therefore the line $x - 2y = 0$ is closed under addition. The line is also closed under scalar multiplication since

$$\alpha(x - 2y) = 0$$

for any α.

EXAMPLE 5-3
Show that the set of second-order polynomials

$$a x^2 + b x + c$$

is a vector space.

SOLUTION 5-3
We denote two vectors in the space by $u = a_2x^2 + a_1x + a_0$ and $v = b_2x^2 + b_1x + b_0$. The vectors add as follows:

$$
\begin{aligned}
u + v &= \left(a_2x^2 + a_1x + a_0\right) + \left(b_2x^2 + b_1x + b_0\right) \\
&= a_2x^2 + b_2x^2 + a_1x + b_1x + a_0 + b_0 \\
&= (a_2 + b_2)x^2 + (a_1 + b_1)x + (a_0 + b_0)
\end{aligned}
$$

The result is another second-order polynomial; therefore, the space is closed under addition. We see immediately that closure under scalar multiplication is also satisfied, since given any scalar α we have

$$\alpha u = \alpha \left(a_2x^2 + a_1x + a_0\right) = (\alpha a_2)x^2 + (\alpha a_1)x + \alpha a_0$$

The result is another second-order polynomial; therefore, the space is closed under scalar multiplication (see Fig. 5-3). There exists a zero vector for this space, which is found by setting $a_2 = a_1 = a_0 = 0$ and so clearly

$$0 + v = (0)x^2 + (0)x + 0 + b_2x^2 + b_1x + b_0 = b_2x^2 + b_1x + b_0$$
$$= v = v + 0$$

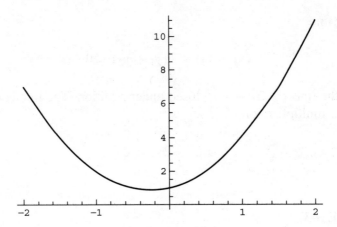

Fig. 5-3. Polynomials can be thought of as a vector space.

There exists an additive inverse of u found by setting

$$a_2 \rightarrow -a_2$$

$$a_1 \rightarrow -a_1$$

$$a_0 \rightarrow -a_0$$

So we have

$$u + (-u) = a_2 x^2 + a_1 x + a_0 + \left(-a_2 x^2 - a_1 x - a_0\right)$$

$$= (a_2 - a_2)\, x^2 + (a_1 - a_1)\, x + a_0 - a_0 = 0$$

EXAMPLE 5-4

Describe the vector space $\mathbb{C}^3$, a three-dimensional complex vector space.

SOLUTION 5-4

$\mathbb{C}^3$ is a vector space over the complex numbers consisting of three-dimensional n-tuples. A vector in this space is a list of three complex numbers of the form

$$a = (a_1, a_2, a_3)$$

Vector addition is carried out componentwise, giving a new list of three complex numbers:

$$a + b = (a_1, a_2, a_3) + (b_1, b_2, b_3) = (a_1 + b_1, a_2 + b_2, a_3 + b_3)$$

Hence this space is closed under addition. Scalar multiplication proceeds in the following way:

$$\alpha a = \alpha\,(a_1, a_2, a_3) = (\alpha a_1, \alpha a_2, \alpha a_3)$$

Since the result is a new listing of three complex numbers, the space is closed under scalar multiplication. The zero vector in $\mathbb{C}^3$ is a list of three zeros:

$$0 = (0, 0, 0)$$

and the inverse of a vector a is given by

$$-a = (-a_1, -a_2, -a_3)$$

EXAMPLE 5-5
The set of functions $f(x)$ into the real numbers is a vector space. Vector addition in this space is defined by the addition of two functions:

$$(f + g)(x) = f(x) + g(x)$$

Scalar multiplication of a function $f(x)$ is given by the product of a scalar $\alpha \in R$ defined as

$$(\alpha f)(x) = \alpha f(x)$$

The zero vector maps every x into 0, i.e.,

$$0(x) = 0 \quad \forall x$$

and the inverse of a vector in this space is the negative of the function:

$$(-f)(x) = -f(x)$$

Basis Vectors

Given a vector u that belongs to a vector space V, we can write u as a *linear combination* of vectors $v_1, v_2, \ldots, v_n$ if there exist scalars $\alpha_1, \alpha_2, \ldots, \alpha_n$ such that

$$u = \alpha_1 v_1 + \alpha_2 v_2 + \cdots + \alpha_n v_n$$

EXAMPLE 5-6

Consider the three-dimensional vector space $\mathbb{C}^3$. Show that we can write the vector

$$u = (2i, 1 + i, 3)$$

as a linear combination of the set $e_1 = (1, 0, 0)$, $e_2 = (0, 1, 0)$, $e_3 = (0, 0, 1)$.

SOLUTION 5-6

Considering the set e_i first, we use the rules of vector addition and scalar multiplication to change the way the vector is written. First, since

$$u + v = (u_1 + v_1, u_2 + v_2, u_3 + v_3)$$

we can rewrite the vector as

$$u = (2i, 1 + i, 3) = (2i, 0, 0) + (0, 1 + i, 3)$$
$$= (2i, 0, 0) + (0, 1 + i, 0) + (0, 0, 3)$$

Now, the rule for scalar multiplication is

$$\alpha u = \alpha (u_1, u_2, u_3) = (\alpha u_1, \alpha u_2, \alpha u_3)$$

This allows us to pull out the factors in each term, i.e.,

$$(2i, 0, 0) = 2i (1, 0, 0) = 2i e_1$$
$$(0, 1 + i, 0) = (1 + i)(0, 1, 0) = (1 + i) e_2$$
$$(0, 0, 3) = 3 (0, 0, 1) = 3 e_3$$

and so we have found that

$$u = 2i\,e_1 + (1+i)\,e_2 + 3\,e_3$$

EXAMPLE 5-7
Write the polynomial

$$u = 4x^2 - 2x + 5$$

as a linear combination of the polynomials

$$p_1 = 2x^2 + x + 1, \qquad p_2 = x^2 - 2x + 2, \qquad p_3 = x^2 + 3x + 6$$

SOLUTION 5-7
If u can be written as a linear combination of these polynomials, then there exist scalars a, b, c such that

$$u = ap_1 + bp_2 + cp_3$$

Using the rules of vector addition and scalar multiplication, we have

$$
\begin{aligned}
u &= ap_1 + bp_2 + cp_3 \\
&= a\left(2x^2 + x + 1\right) + b\left(x^2 - 2x + 2\right) + c\left(x^2 + 3x + 6\right) \\
&= (2a + b + c)\,x^2 + (a - 2b + 3c)\,x + (a + 2b + 6c)
\end{aligned}
$$

Comparison with $u = 4x^2 - 2x + 5$ yields three equations

$$
\begin{aligned}
2a + b + c &= 4 \\
a - 2b + 3c &= -2 \\
a + 2b + 6c &= 5
\end{aligned}
$$

This is a linear system in (a, b, c) that can be represented with the augmented matrix

$$
\begin{bmatrix}
2 & 1 & 1 & 4 \\
1 & -2 & 3 & -2 \\
1 & 2 & 6 & 5
\end{bmatrix}
$$

The operations $R_1 - 2R_2 \to R_2$ and $R_1 - 2R_3 \to R_3$ yield

$$\begin{bmatrix} 2 & 1 & 1 & | & 4 \\ 0 & 5 & -5 & | & 8 \\ 0 & -3 & -11 & | & -6 \end{bmatrix}$$

$3R_2 + 5R_3 \to R_3$ gives

$$\begin{bmatrix} 2 & 1 & 1 & | & 4 \\ 0 & 5 & -5 & | & 8 \\ 0 & 0 & -70 & | & -6 \end{bmatrix}$$

The last row yields

$$c = \frac{3}{35}$$

Back substitution into the second row gives

$$b = c + \frac{8}{5} = \frac{3}{35} + \frac{8}{5} = \frac{56}{35}$$

The first row then allows us to solve for a

$$a = 2 - \frac{1}{2}b - \frac{1}{2}c = 2 - \frac{56}{70} - \frac{3}{70} = \frac{81}{70}$$

Therefore, we can write $u = 4x^2 - 2x + 5$ in terms of the polynomials p_1, p_2, p_3 as

$$u = \frac{81}{70}p_1 + \frac{56}{35}p_2 + \frac{3}{35}p_3$$

A SPANNING SET

A set of vectors $\{u_1, u_2, \ldots, u_n\}$ is said to span a vector space V if every vector $v \in V$ can be written as a linear combination of $\{u_1, u_2, \ldots, u_n\}$; in other words, we can write *any* vector v in the space as a linear combination

$$v = \alpha_1 u_1 + \alpha_2 u_2 + \cdots + \alpha_n u_n$$

for scalars α_i.

EXAMPLE 5-8
We have already seen the set $e_1 = (1, 0, 0)$, $e_2 = (0, 1, 0)$, $e_3 = (0, 0, 1)$. Any vector in $\mathbb{C}^3$ can be written in terms of this set, since

$$(\alpha, \beta, \gamma) = \alpha (1, 0, 0) + \beta (0, 1, 0) + \gamma (0, 0, 1)$$

for any complex numbers α, β, γ. Therefore $e_1 = (1, 0, 0)$, $e_2 = (0, 1, 0)$, $e_3 = (0, 0, 1)$ span $\mathbb{C}^3$.

Linear Independence

A collection of vectors $\{u_1, u_2, \ldots, u_n\}$ is *linearly independent* if the equation

$$\alpha_1 u_1 + \alpha_2 u_2 + \cdots + \alpha_n u_n = 0$$

implies that $\alpha_1 = \alpha_2 = \cdots = \alpha_n = 0$. If this condition is not met then we say that the set of vectors $\{u_1, u_2, \ldots, u_n\}$ is linearly dependent. Said another way, if a set of vectors is linearly independent, then no vector from the set can be written as a linear combination of the other vectors.

EXAMPLE 5-9
Show that the set

$$a = (1, 2, 1), \qquad b = (0, 1, 0), \qquad c = (-2, 0, -2)$$

is linearly dependent.

SOLUTION 5-9
We can write

$$2b - \frac{1}{2}c = 2 (0, 1, 0) - \frac{1}{2}(-2, 0, -2) = (0, 2, 0) + (1, 0, 1)$$
$$= (1, 2, 1) = a$$

Since a can be written as a linear combination of b, c, the set is linearly dependent.

EXAMPLE 5-10
Show that the set

$$(1, 0, 1), (1, 1, -1), (0, 1, 0)$$

is linearly independent.

SOLUTION 5-10

Given scalars a, b, c we have

$$a(1, 0, 1) + b(1, 1, -1) + c(0, 1, 0) = (a + b, b + c, a - b)$$

The zero vector is

$$(0, 0, 0)$$

Therefore to have

$$a(1, 0, 1) + b(1, 1, -1) + c(0, 1, 0) = 0$$

It must be the case that

$$a + b = 0$$
$$b + c = 0$$
$$a - b = 0$$

From the third equation we see that $a = b$. Substitution into the first equation yields

$$a + b = a + a = 2a = 0$$
$$\Rightarrow a = 0$$

From this we conclude that $b = c = 0$ as well. Since all of the constants are zero, the set is linearly independent.

We can show that a set of vectors is linearly independent by arranging them in a matrix form. Then row reduce the matrix; if each row has a nonzero pivot, then the vectors are linearly independent.

EXAMPLE 5-11

Determine if the set $\{(1, 3, 5), (4, -1, 2), (0, -1, 2)\}$ is linearly independent.

SOLUTION 5-11
We arrange each set as a matrix, using each vector as a column. For the first set $\{(1, 3, 5), (4, -1, 2), (0, -1, 2)\}$ the matrix is

$$A = \begin{bmatrix} 1 & 4 & 0 \\ 3 & -1 & -1 \\ 5 & 2 & 2 \end{bmatrix}$$

Now we row reduce the matrix

$$\begin{bmatrix} 1 & 4 & 0 \\ 3 & -1 & -1 \\ 5 & 2 & 2 \end{bmatrix} \sim \begin{bmatrix} 1 & 4 & 0 \\ 0 & -13 & -1 \\ 5 & 2 & 2 \end{bmatrix} \sim \begin{bmatrix} 1 & 4 & 0 \\ 0 & -13 & -1 \\ 0 & -18 & 2 \end{bmatrix} \sim \begin{bmatrix} 1 & 4 & 0 \\ 0 & -13 & -1 \\ 0 & 0 & 44 \end{bmatrix}$$

The operations used were $-3R_1 + R_2 \rightarrow R_2, -5R_1 + R_3 \rightarrow R_3$, and $-18R_2 + 13R_3 \rightarrow R_3$. Since all the columns in the reduced matrix contain a pivot entry, no vector can be written as a linear combination of the other vectors; therefore, the set is linearly independent.

EXAMPLE 5-12
Does the set $(1, 1, 1, 1), (1, 3, 2, 1), (2, 3, 6, 4), (2, 2, 2, 2)$ span $\mathbb{R}^4$?

SOLUTION 5-12
We arrange the set in matrix form:

$$A = \begin{bmatrix} 1 & 1 & 1 & 1 \\ 1 & 3 & 2 & 1 \\ 2 & 3 & 6 & 4 \\ 2 & 2 & 2 & 2 \end{bmatrix}$$

Next we row reduce the matrix:

$$\begin{bmatrix} 1 & 1 & 1 & 1 \\ 1 & 3 & 2 & 1 \\ 2 & 3 & 6 & 4 \\ 2 & 2 & 2 & 2 \end{bmatrix} \sim \begin{bmatrix} 1 & 1 & 1 & 1 \\ 0 & 2 & 1 & 0 \\ 2 & 3 & 6 & 4 \\ 2 & 2 & 2 & 2 \end{bmatrix} \sim \begin{bmatrix} 1 & 1 & 1 & 1 \\ 0 & 2 & 1 & 0 \\ 0 & 1 & 4 & 2 \\ 2 & 2 & 2 & 2 \end{bmatrix} \sim \begin{bmatrix} 1 & 1 & 1 & 1 \\ 0 & 2 & 1 & 0 \\ 0 & 1 & 4 & 2 \\ 0 & 0 & 0 & 0 \end{bmatrix}$$

Since the last row is all zeros, this set of vectors is linearly dependent. Therefore they cannot form a basis of $\mathbb{R}^4$. The echelon matrix has three nonzero rows. Therefore the set spans a subspace of dimension 3.

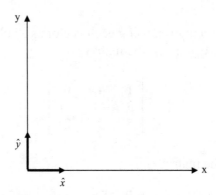

Fig. 5-4. The basis set $\{\hat{x}, \hat{y}\}$ spans the $x-y$ plane. But we could equally well use the basis vectors $\{\hat{r}, \hat{\phi}\}$ to write any vector in plane polar coordinates.

Basis Vectors

If a set of vectors $\{u_1, u_2, \ldots, u_n\}$ spans a vector space V and is linearly independent, we say that this set is a *basis* of V. Any vector that belongs to V can be written as a unique linear combination of the basis $\{u_1, u_2, \ldots, u_n\}$. There exist multiple bases for a given vector space V; in fact there can be infinitely many (see Fig. 5-4).

$$\hat{x}\,\hat{y}$$

EXAMPLE 5-13
The set

$$e_1 = (1, 0, 0), \qquad e_2 = (0, 1, 0), \qquad e_3 = (0, 0, 1)$$

is a basis for the vector space R^3.

Completeness

Completeness or the *closure relation* means that we can write the identity in terms of outer products of a set of basis vectors. An outer product is a matrix multiplication operation between a column vector and a row vector.

The result of an outer product is a matrix, calculated by

$$\begin{pmatrix} a_1 \\ a_2 \\ \vdots \\ a_n \end{pmatrix} (b_1 \quad b_2 \quad \cdots \quad b_n) = \begin{bmatrix} a_1b_1 & a_1b_2 & \cdots & a_1b_n \\ a_2b_1 & a_2b_2 & \cdots & a_2b_n \\ \vdots & \vdots & \ddots & \vdots \\ a_nb_1 & a_nb_2 & \cdots & a_nb_n \end{bmatrix}$$

EXAMPLE 5-14
Find the outer product of

$$(1 \quad 2 \quad 3), (4 \quad 5 \quad 6)$$

SOLUTION 5-14
The outer product is

$$\begin{pmatrix} 1 \\ 2 \\ 3 \end{pmatrix} (4 \quad 5 \quad 6) = \begin{bmatrix} (1)(4) & (1)(5) & (1)(6) \\ (2)(4) & (2)(5) & (2)(6) \\ (3)(4) & (3)(5) & (3)(6) \end{bmatrix} = \begin{bmatrix} 4 & 5 & 6 \\ 8 & 10 & 12 \\ 12 & 15 & 18 \end{bmatrix}$$

EXAMPLE 5-15
Show that the set $e_1 = (1, 0, 0)$, $e_2 = (0, 1, 0)$, $e_3 = (0, 0, 1)$ is complete.

SOLUTION 5-15
The identity matrix in 3 dimensions is

$$I_3 = \begin{bmatrix} 1 & 0 & 0 \\ 0 & 1 & 0 \\ 0 & 0 & 1 \end{bmatrix}$$

The first outer product is

$$\begin{pmatrix} 1 \\ 0 \\ 0 \end{pmatrix} (1 \quad 0 \quad 0) = \begin{bmatrix} 1 & 0 & 0 \\ 0 & 0 & 0 \\ 0 & 0 & 0 \end{bmatrix}$$

The second is

$$\begin{pmatrix} 0 \\ 1 \\ 0 \end{pmatrix} (0 \quad 1 \quad 0) = \begin{bmatrix} 0 & 0 & 0 \\ 0 & 1 & 0 \\ 0 & 0 & 0 \end{bmatrix}$$

and the third is

$$
\begin{pmatrix} 0 \\ 0 \\ 1 \end{pmatrix} (0 \quad 0 \quad 1) = \begin{bmatrix} 0 & 0 & 0 \\ 0 & 0 & 0 \\ 0 & 0 & 1 \end{bmatrix}
$$

Summing we obtain the identity matrix, showing the set is complete:

$$
\begin{bmatrix} 1 & 0 & 0 \\ 0 & 0 & 0 \\ 0 & 0 & 0 \end{bmatrix} + \begin{bmatrix} 0 & 0 & 0 \\ 0 & 1 & 0 \\ 0 & 0 & 0 \end{bmatrix} + \begin{bmatrix} 0 & 0 & 0 \\ 0 & 0 & 0 \\ 0 & 0 & 1 \end{bmatrix} = \begin{bmatrix} 1 & 0 & 0 \\ 0 & 1 & 0 \\ 0 & 0 & 1 \end{bmatrix}
$$

DIMENSION OF A VECTOR SPACE

The *dimension* of a vector space n is the minimum number of basis vectors $\{u_1, u_2, \ldots, u_n\}$ required to span the space. If V is a vector space and $\{u_1, u_2, \ldots, u_n\}$ is a basis with n elements and $\{v_1, v_2, \ldots, v_m\}$ is another basis with m elements, then $m = n$. This means that all basis sets of a vector space contain the same number of elements. A vector space that does not have a finite basis is called *infinite dimensional.*

Subspaces

Suppose that V is a vector space. A subset W of V is a *subspace* if W is also a vector space. In other words, closure under vector addition and scalar multiplication must be satisfied for W in order for it to be a subspace.

It is easy to determine if W is a subspace because most of the vector axioms carry over to W automatically. We can verify that W is a subspace by

- Confirming that W has a zero vector.
- Verifying that if $u, v \in W$, then $\alpha u + \beta v \in W$.

EXAMPLE 5-16
Let V be the complex vector space $\mathbb{C}^3$ and let W be the set of vectors for which the third component is zero:

$$
u = (\alpha, \beta, 0) \in W
$$

Is W a subspace of V?

SOLUTION 5-16

For the zero vector, we set $\alpha = \beta = 0$ and obtain

$$0 = (0, 0, 0)$$

Clearly

$$0 + u = (0, 0, 0) + (\alpha, \beta, 0) = (\alpha, \beta, 0) = u$$

for any $u \in W$. Now consider a second element that belongs to W:

$$v = (\gamma, \delta, 0)$$

Let a and b be two scalars, then the linear combination

$$au + bv = a(\alpha, \beta, 0) + b(\gamma, \delta, 0) = (a\alpha, a\beta, 0) + (b\gamma, b\delta, 0)$$
$$= (a\alpha + b\gamma, a\beta + b\delta, 0)$$

We have found that $a u + b v$ is a complex 3-tuple with the third element equal to zero, and therefore $a u + b v \in W$. Both criteria are satisfied and so we conclude that W is a subspace of V.

Row Space of a Matrix

The rows of a matrix A can be viewed as vectors that span a subspace. If the matrix A is a matrix of real numbers then the rows of A span a subspace of $\mathbb{R}^n$, while if A is a matrix of complex numbers the rows of A span a subspace of $\mathbb{C}^n$.

The columns of A can also be viewed as vectors and they form a subspace of $\mathbb{R}^n$ or $\mathbb{C}^n$ in an analogous manner. The following relationship holds:

$$\text{colsp}(A) = \text{rowsp}(A^T)$$

Matrices that are row equivalent, that is, matrices that can be obtained from one another by applying a sequence of elementary row operations, have the same row space. The nonzero rows of an echelon matrix are linearly independent.

To find the row and column spaces of a matrix A

- Reduce the matrix to row echelon form.
- The columns of the row echelon form of the matrix with nonzero pivots identify the *basic columns* of the matrix A.

- The basic columns of A span the column space of A.
- The nonzero rows of the row echelon form of A span the row space of A.

Notice that we must use the basic columns of the original matrix as the basis of the column space of A. Do not use the columns of the echelon matrix.

The *row rank* of a matrix A is the number of vectors needed to span the row space. The *column rank* is the number of vectors needed to span the column space. These values are equal to each other. We can also find the rank of A by adding up the number of leading 1s in the reduced row echelon form of A.

EXAMPLE 5-17

Determine the spanning sets for the row and column spaces of the matrix

$$A = \begin{bmatrix} 1 & 2 & 1 & 3 \\ -2 & -1 & 3 & 5 \\ 3 & 4 & 3 & -1 \end{bmatrix}$$

SOLUTION 5-17

We begin by applying $2R_1 + R_2 \rightarrow R_2$ and obtain

$$A = \begin{bmatrix} 1 & 2 & 1 & 3 \\ -2 & -1 & 3 & 5 \\ 3 & 4 & 3 & -1 \end{bmatrix} \sim \begin{bmatrix} 1 & 2 & 1 & 3 \\ 0 & 3 & 5 & 11 \\ 3 & 4 & 3 & -1 \end{bmatrix}$$

Next we use $-3R_1 + R_3 \rightarrow R_3$:

$$\begin{bmatrix} 1 & 2 & 1 & 3 \\ 0 & 3 & 5 & 11 \\ 3 & 4 & 3 & -1 \end{bmatrix} \sim \begin{bmatrix} 1 & 2 & 1 & 3 \\ 0 & 3 & 5 & 11 \\ 0 & -2 & 0 & -10 \end{bmatrix}$$

Now take $2R_2 + 3R_3 \rightarrow R_3$, which gives

$$\begin{bmatrix} 1 & 2 & 1 & 3 \\ 0 & 3 & 5 & 11 \\ 0 & 0 & 10 & -8 \end{bmatrix}$$

We divide the last row by 2:

$$\begin{bmatrix} 1 & 2 & 1 & 3 \\ 0 & 3 & 5 & 11 \\ 0 & 0 & 5 & -4 \end{bmatrix}$$

Then we take $-R_3 + R_2 \to R_2$ and then divide the second row by 3 and find

$$\begin{bmatrix} 1 & 2 & 1 & 3 \\ 0 & 3 & 0 & 15 \\ 0 & 0 & 5 & -4 \end{bmatrix} \sim \begin{bmatrix} 1 & 2 & 1 & 3 \\ 0 & 1 & 0 & 5 \\ 0 & 0 & 5 & -4 \end{bmatrix}$$

Next we take $R_3 - 5R_1 \to R_1$ and then divide the third row by 5:

$$\begin{bmatrix} 1 & 2 & 1 & 3 \\ 0 & 1 & 0 & 5 \\ 0 & 0 & 5 & -4 \end{bmatrix} \sim \begin{bmatrix} 1 & 2 & 0 & -19 \\ 0 & 1 & 0 & 5 \\ 0 & 0 & 5 & -4 \end{bmatrix} \sim \begin{bmatrix} 1 & 2 & 0 & -19 \\ 0 & 1 & 0 & 5 \\ 0 & 0 & 1 & -4/5 \end{bmatrix}$$

Finally, we use $-2R_2 + R_1 \to R_1$:

$$\begin{bmatrix} 1 & 2 & 0 & -19 \\ 0 & 1 & 0 & 5 \\ 0 & 0 & 1 & -4/5 \end{bmatrix} \sim \begin{bmatrix} 1 & 0 & 0 & -29 \\ 0 & 1 & 0 & 5 \\ 0 & 0 & 1 & -4/5 \end{bmatrix}$$

There are three nonzero rows; therefore, the row space is spanned by

$$\text{rowsp}(A) = \left\{ \begin{pmatrix} 1 \\ 0 \\ 0 \\ -29 \end{pmatrix}, \begin{pmatrix} 0 \\ 1 \\ 0 \\ 5 \end{pmatrix}, \begin{pmatrix} 0 \\ 0 \\ 1 \\ -\frac{4}{5} \end{pmatrix} \right\}$$

To find the column space of A, first we identify the columns that contain pivots in the row echelon form of A. These are the first and second columns. We underline the pivots

$$\begin{bmatrix} \underline{1} & 0 & 0 & -29 \\ 0 & \underline{1} & 0 & 5 \\ 0 & 0 & \underline{1} & -4/5 \end{bmatrix}$$

There are three leading 1s in the reduced form and so the rank of the matrix is 3. The vectors that span the column space are from the corresponding columns of A:

$$A = \begin{bmatrix} 1 & 2 & 1 & 3 \\ -2 & -1 & 3 & 5 \\ 3 & 4 & 3 & -1 \end{bmatrix}$$

and so

$$\text{colsp}(A) = \left\{ \begin{pmatrix} 1 \\ -2 \\ 3 \end{pmatrix}, \begin{pmatrix} 2 \\ -1 \\ 4 \end{pmatrix}, \begin{pmatrix} 1 \\ 3 \\ 3 \end{pmatrix} \right\}$$

Notice that the row rank = 3 = column rank of A, since three vectors are needed to span the row and column spaces of the matrix.

EXAMPLE 5-18
Find the row and column spaces of

$$A = \begin{bmatrix} 1 & 2 & 3 \\ 4 & 5 & 6 \\ 7 & 8 & 9 \end{bmatrix}$$

SOLUTION 5-18
We row reduce the matrix. First take $-4R_1 + R_2 \rightarrow R_2$:

$$A = \begin{bmatrix} 1 & 2 & 3 \\ 4 & 5 & 6 \\ 7 & 8 & 9 \end{bmatrix} \sim \begin{bmatrix} 1 & 2 & 3 \\ 0 & -3 & -6 \\ 7 & 8 & 9 \end{bmatrix}$$

We eliminate the first term in the third row with $-7R_1 + R_3 \rightarrow R_3$:

$$\begin{bmatrix} 1 & 2 & 3 \\ 0 & -3 & -6 \\ 7 & 8 & 9 \end{bmatrix} \sim \begin{bmatrix} 1 & 2 & 3 \\ 0 & -3 & -6 \\ 0 & -6 & -12 \end{bmatrix}$$

Now divide row 2 by -3, and row 3 by -6:

$$\begin{bmatrix} 1 & 2 & 3 \\ 0 & -3 & -6 \\ 0 & -6 & -12 \end{bmatrix} \sim \begin{bmatrix} 1 & 2 & 3 \\ 0 & 1 & 2 \\ 0 & 1 & 2 \end{bmatrix}$$

We can eliminate the third row with $-R_2 + R_3 \rightarrow R_3$:

$$\begin{bmatrix} 1 & 2 & 3 \\ 0 & 1 & 2 \\ 0 & 1 & 2 \end{bmatrix} \sim \begin{bmatrix} 1 & 2 & 3 \\ 0 & 1 & 2 \\ 0 & 0 & 0 \end{bmatrix}$$

We finish with $-2R_2 + R_1 \to R_1$:

$$\begin{bmatrix} 1 & 2 & 3 \\ 0 & 1 & 2 \\ 0 & 0 & 0 \end{bmatrix} \sim \begin{bmatrix} 1 & 0 & -1 \\ 0 & 1 & 2 \\ 0 & 0 & 0 \end{bmatrix}$$

The row space is given by the nonzero rows of the reduced matrix. Therefore,

$$\text{rowsp}(A) = \left\{ \begin{pmatrix} 1 \\ 0 \\ -1 \end{pmatrix}, \begin{pmatrix} 0 \\ 1 \\ 2 \end{pmatrix} \right\}$$

The nonzero pivots are underlined here:

$$\begin{bmatrix} \underline{1} & 0 & -1 \\ 0 & \underline{1} & 2 \\ 0 & 0 & 0 \end{bmatrix}$$

So the first two columns of A span the column space. These are

$$\text{colsp}(A) = \left\{ \begin{pmatrix} 1 \\ 4 \\ 7 \end{pmatrix}, \begin{pmatrix} 2 \\ 5 \\ 8 \end{pmatrix} \right\}$$

We have

$$\text{rowrank}\,(A) = 2 = \text{colrank}\,(A)$$

Also notice that in the reduced echelon form of the matrix, there are two leading 1s and so the rank of the matrix is 2.

EXAMPLE 5-19
Find the row and column spaces of

$$A = \begin{bmatrix} 1 & 2 & -4 & 3 & -1 \\ 1 & 2 & -2 & 2 & 1 \\ 2 & 4 & -2 & 3 & 4 \end{bmatrix}$$

SOLUTION 5-19

We row reduce the matrix with the following steps:

$$-R_1 + R_2 \rightarrow R_2, -2R_1 + R_3 \rightarrow R_3, -3R_2 + R_3 \rightarrow R_3$$

This results in

$$A = \begin{bmatrix} 1 & 2 & -4 & 3 & -1 \\ 1 & 2 & -2 & 2 & 1 \\ 2 & 4 & -2 & 3 & 4 \end{bmatrix} \sim \begin{bmatrix} 1 & 2 & -4 & 3 & -1 \\ 0 & 0 & 2 & -1 & 2 \\ 2 & 4 & -2 & 3 & 4 \end{bmatrix}$$

$$\sim \begin{bmatrix} 1 & 2 & -4 & 3 & -1 \\ 0 & 0 & 2 & -1 & 2 \\ 0 & 0 & 6 & -3 & 6 \end{bmatrix} \sim \begin{bmatrix} 1 & 2 & -4 & 3 & -1 \\ 0 & 0 & 2 & -1 & 2 \\ 0 & 0 & 0 & 0 & 0 \end{bmatrix}$$

There are two nonzero rows in the echelon matrix. So the row space is

$$\text{rowsp}(A) = \left\{ \begin{pmatrix} 1 \\ 2 \\ -4 \\ 3 \\ -1 \end{pmatrix}, \begin{pmatrix} 0 \\ 0 \\ 2 \\ -1 \\ 2 \end{pmatrix} \right\}$$

The pivots in the echelon matrix are underlined:

$$\begin{bmatrix} \underline{1} & 2 & -4 & 3 & -1 \\ 0 & 0 & \underline{2} & -1 & 2 \\ 0 & 0 & 0 & 0 & 0 \end{bmatrix}$$

So the first and third columns of A span the column space:

$$\text{colsp}(A) = \left\{ \begin{pmatrix} 1 \\ 1 \\ 2 \end{pmatrix}, \begin{pmatrix} -4 \\ -2 \\ -2 \end{pmatrix} \right\}$$

Notice that two vectors are required to span the row and column spaces of the matrix. Therefore, the rank of A is 2.

Null Space of a Matrix

The null space of a matrix is found from

$$Ax = 0$$

where the set of vectors x is the basis of the null space of A. *Nullity* is the number of parameters needed in the solution to this equation. When the matrix A is $m \times n$ then

$$\text{rank}(A) + \text{nullity}(A) = n$$

From this relation we see that the null space of a matrix is 0 if the rank $= n$. The best way to explain how to find the null space of a matrix is with examples.

EXAMPLE 5-20
Find the null space for

$$A = \begin{bmatrix} 1 & -1 & 2 \\ -1 & 1 & -2 \end{bmatrix}$$

SOLUTION 5-20
We immediately reduce the matrix to

$$A = \begin{bmatrix} 1 & -1 & 2 \\ -1 & 1 & -2 \end{bmatrix} \sim \begin{bmatrix} 1 & -1 & 2 \\ 0 & 0 & 0 \end{bmatrix}$$

We find the null space of the matrix from the solution of $Ax = 0$ for a vector x:

$$x = \begin{pmatrix} x_1 \\ x_2 \\ x_3 \end{pmatrix}$$

The reduced form of the matrix gives the equation

$$x_1 - x_2 + 2x_3 = 0$$

Solving this equation, we get

$$x_1 = x_2 - 2x_3$$

So the solution is a vector of the form

$$\begin{pmatrix} x_1 \\ x_2 \\ x_3 \end{pmatrix} = \begin{pmatrix} x_2 - 2x_3 \\ x_2 \\ x_3 \end{pmatrix}$$

To find the null space, we rewrite this vector so that x_2 and x_3 are coefficients multiplying two vectors:

$$\begin{pmatrix} x_2 - 2x_3 \\ x_2 \\ x_3 \end{pmatrix} = x_2 \begin{pmatrix} 1 \\ 1 \\ 0 \end{pmatrix} + x_3 \begin{pmatrix} -2 \\ 0 \\ 1 \end{pmatrix}$$

The null space of A is the set of all linear combinations of the vectors

$$h_1 = \begin{pmatrix} 1 \\ 1 \\ 0 \end{pmatrix}, \qquad h_2 = \begin{pmatrix} -2 \\ 0 \\ 1 \end{pmatrix}$$

EXAMPLE 5-21
Find a basis for the null space of the matrix

$$A = \begin{bmatrix} 1 & 2 & -4 & 3 & -1 \\ 1 & 2 & -2 & 2 & 1 \\ 2 & 4 & -2 & 3 & 4 \end{bmatrix}$$

SOLUTION 5-21
The solution to $Ax = 0$ is a vector $(x_1, x_2, x_3, x_4, x_5)$. As usual, the first step is to reduce the matrix. In the example above we found that the reduced form of this matrix is

$$\begin{bmatrix} 1 & 2 & -4 & 3 & -1 \\ 0 & 0 & 2 & -1 & 2 \\ 0 & 0 & 0 & 0 & 0 \end{bmatrix}$$

The pivots are located in column 1 and column 3. Columns that do not have pivots tell us *free variables* or parameters for the matrix. In this case the free columns are 2, 4, and 5. Therefore the free variables are

$$(x_2, x_4, x_5)$$

We write the variables (x_1, x_3) in terms of the free variables using the equations represented by the reduced form of the matrix. From the first row, we have

$$x_1 + 2x_2 - 4x_3 + 3x_4 - x_5 = 0$$

The second row tells us

$$2x_3 - x_4 + 2x_5 = 0$$

First we rearrange the second equation to obtain

$$x_3 = \frac{1}{2}x_4 - x_5$$

This allows us to simplify the first equation

$$x_1 = -2x_2 + 4x_3 - 3x_4 + x_5 = -2x_2 - x_4 - 3x_5$$

Therefore we have

$$
\begin{pmatrix} x_1 \\ x_2 \\ x_3 \\ x_4 \\ x_5 \end{pmatrix} =
\begin{pmatrix} -2x_2 - x_4 - 3x_5 \\ x_2 \\ \frac{1}{2}x_4 - x_5 \\ x_4 \\ x_5 \end{pmatrix} =
x_2 \begin{pmatrix} -2 \\ 1 \\ 0 \\ 0 \\ 0 \end{pmatrix} +
x_4 \begin{pmatrix} -1 \\ 0 \\ \frac{1}{2} \\ 1 \\ 0 \end{pmatrix} +
x_5 \begin{pmatrix} -3 \\ 0 \\ -1 \\ 0 \\ 1 \end{pmatrix}
$$

The null space of A is given by the set of all linear combinations of the vectors

$$
h_1 = \begin{pmatrix} -2 \\ 1 \\ 0 \\ 0 \\ 0 \end{pmatrix}, \qquad
h_2 = \begin{pmatrix} -1 \\ 0 \\ \frac{1}{2} \\ 1 \\ 0 \end{pmatrix}, \qquad
h_3 = \begin{pmatrix} -3 \\ 0 \\ -1 \\ 0 \\ 1 \end{pmatrix}
$$

Quiz

1. Is the line $3x + 5y = 2$ a vector space? If not, why not?
2. Is the set of vectors $A_x \hat{x} + A_y \hat{y} + 2\hat{z}$ a vector space? If not, why not?
3. Show that the set of second-order polynomials is commutative and associative under addition, is associative and distributive under scalar multiplication, and that there exists an identity.

4. Show that the set of 2-tuples of real numbers

$$\begin{bmatrix} \alpha \\ \beta \end{bmatrix}$$

is a vector space.

5. Write the vector

$$u = (2i, 1 + i, 3)$$

as a linear combination of the set $v_1 = (1, 1, 1)$, $v_2 = (1, 0, -1)$, $v_3 = (1, -1, 1)$.

6. Write the polynomial

$$v = 5t^2 - 4t + 1$$

as a linear combination of the polynomials

$$p_1 = 2t^2 + 9t - 1, \qquad p_2 = 4t + 2, \qquad p_3 = t^2 + 3t + 6$$

7. Consider the set of 2×2 matrix of complex numbers

$$\begin{bmatrix} \alpha & \beta \\ \gamma & \delta \end{bmatrix}$$

Show that this matrix is a vector space. Find a set of matrix that spans the space.

8. Is the set $(-2, 1, 1)$, $(4, 0, 0)$, $(0, 2, 0)$ linearly independent?

9. Is the set

$$v_1 = \frac{1}{\sqrt{2}} \begin{pmatrix} 1 \\ 1 \end{pmatrix}, \qquad v_2 = \frac{1}{\sqrt{2}} \begin{pmatrix} 1 \\ -1 \end{pmatrix}$$

complete?

10. Is the set W of vectors of the form (a, b, c), where $a = b = c$, a subspace of $\mathbb{R}^3$?

11. Find the row space and column space of

$$A = \begin{bmatrix} 1 & -1 & 2 & 5 \\ 2 & 4 & 1 & 0 \\ -1 & 3 & 0 & 1 \end{bmatrix}$$

12. Find the null space of

$$A = \begin{bmatrix} 1 & 2 & 3 \\ 4 & 5 & 6 \\ 7 & 8 & 9 \end{bmatrix}$$

13. Find the row space, column space, and null space of

$$B = \begin{bmatrix} 1 & -2 & 1 & 0 \\ 3 & 1 & 4 & 5 \\ 2 & 3 & 5 & -1 \end{bmatrix}$$

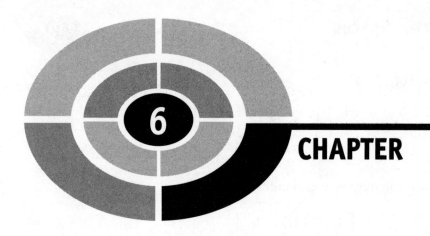

6 CHAPTER

Inner Product Spaces

When we introduced vectors in chapter 4, we briefly discussed the notion of an inner product. In this chapter we will investigate this notion in more detail. We begin with a formal definition.

Let V be a vector space. To each pair of vectors $u, v \in V$ there is a *number* that we denote (u, v) that is called the *inner product,* if it satisfies the following:

1. *Linearity.* For a real vector space, the inner product is a real number and the inner product satisfies

$$(au + bv, w) = a(u, w) + b(v, w)$$

If the vector space is complex, the inner product is a complex number. We will define it in the following way. It is *antilinear* in the first argument

$$(au + bv, w) = a^*(u, w) + b^*(v, w)$$

but is linear in the second argument

$$(u, av + bw) = a(u, v) + b(u, w)$$

2. *Symmetry.* For a real vector space, the inner product is symmetric

$$(u, v) = (v, u)$$

If the vector space is complex, then the inner product is conjugate symmetric

$$(u, v) = (v, u)^*$$

3. *Positive Definiteness.* This means that the inner product satisfies

$$(u, u) \geq 0$$

with equality if and only if $u = 0$.

EXAMPLE 6-1
Suppose that V is a real vector space and that

$$(u, v) = -2$$
$$(u, w) = 5$$

Calculate $(3v - 6w, u)$.

SOLUTION 6-1
First we use the linearity property. The vector space is real, and so we have

$$(3v - 6w, u) = 3(v, u) - 6(w, u)$$

We also know that a real vector space obeys the symmetry property. Therefore we can rewrite this as

$$3(v, u) - 6(w, u) = 3(u, v) - 6(u, w)$$

Now, using the given information, we find

$$(3v - 6w, u) = 3(u, v) - 6(u, w) = (3)(-2) - (6)(5) = -6 - 30 = -36$$

The Vector Space $\mathbb{R}^n$

We define a vector u in $\mathbb{R}^n$ as the n-tuple $(u_1, u_2, \ldots, u_n)$. The inner product for the Euclidean space $\mathbb{R}^n$ is given by

$$(u, v) = u_1 v_1 + u_2 v_2 + \cdots + u_n v_n$$

The norm of a vector is denoted by $\|u\|$ and is calculated using

$$\|u\| = \sqrt{(u, u)} = \sqrt{u_1^2 + u_2^2 + \cdots + u_n^2}$$

EXAMPLE 6-2
Let $u = (-3, 4, 1)$, and $v = (2, 1, 1)$ be vectors in $\mathbb{R}^3$. Find the norm of each vector.

SOLUTION 6-2
Using the formula with $n = 3$, we have

$$\|u\| = \sqrt{(u, u)} = \sqrt{u_1^2 + u_2^2 + u_3^2} = \sqrt{(-3)^2 + (4)^2 + (1)^2}$$
$$= \sqrt{9 + 16 + 1} = \sqrt{27}$$

and

$$\|v\| = \sqrt{(v, v)} = \sqrt{v_1^2 + v_2^2 + v_3^2} = \sqrt{(2)^2 + (1)^2 + (1)^2} = \sqrt{4 + 1 + 1} = \sqrt{6}$$

EXAMPLE 6-3
Suppose that $u = (-1, 3, 2)$, and $v = (2, 0, 1)$ are vectors in $\mathbb{R}^3$. Find the angle between these two vectors.

SOLUTION 6-3
In Chapter 4 we learned that the angle between two vectors can be found from the inner product using

$$\cos \theta = \frac{(u, v)}{\|u\| \, \|v\|}$$

The inner product is

$$(u, v) = (-1)(2) + (3)(0) + (2)(1) = -2 + 0 + 2 = 0$$

Therefore we have

$$\cos \theta = 0$$

Which leads to

$$\theta = \frac{\pi}{2}$$

We have found that this pair of vectors is *orthogonal*. For ordinary vectors in Euclidean space, the vectors are perpendicular, as the calculation of angle shows.

Inner Products on Function Spaces

Looking at the formula for the inner product, one can see that we can generalize this notion to a function by letting summations go to integrals. The vector space $C\,[a,\,b]$ is the space of all continuous functions on the closed interval $a \leq x \leq b$. Supposing that $f(x)$ and $g(x)$ are two functions that belong to $C\,[a,\,b]$, the inner product is given by

$$(f, g) = \int_a^b f(x)\, g(x)\, dx$$

EXAMPLE 6-4
Let $C\,[0,\,1]$ be the function space of polynomials defined on the closed interval $0 \leq x \leq 1$ and let

$$f(x) = -2x + 1, \quad g(x) = 5x^2 - 2x$$

Find the norm of each function and then compute their inner product.

SOLUTION 6-4
First we compute the norm of $f(x) = -2x + 1$, which is shown in Fig. 6-1.
   The norm is given by

$$(f, f) = \int_0^1 f^2(x)\, dx = \int_0^1 (-2x + 1)^2 dx = \int_0^1 (4x^2 - 4x + 1)\, dx$$

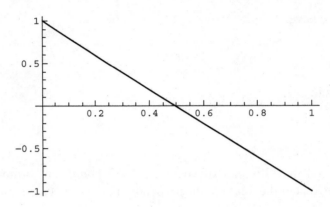

Fig. 6-1. The function $-2x + 1$ which belongs to the vector space $C[0, 1]$.

Integrating term by term, we find

$$(f, f) = \frac{4}{3}x^3 - 2x^2 + x \Big|_0^1 = \frac{4}{3} - 2 + 1 = \frac{1}{3}$$

Now we consider $g(x) = 5x^2 - 2x$. The function is shown in Fig. 6-2.
The norm is given by

$$(g, g) = \int_0^1 g^2(x)\, dx = \int_0^1 \left(5x^2 - 2x\right)^2 dx = \int_0^1 \left(25x^4 - 20x^3 + 4x^2\right) dx$$

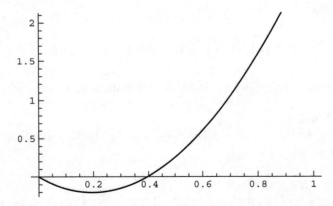

Fig. 6-2. $g(x) = 5x^2 - 2x$ is also continuous over the interval and so belongs to $C[0, 1]$.

Integrating term by term, we obtain

$$(g, g) = 5x^5 - 5x^4 + \frac{4}{3}x^3 \Big|_0^1 = \frac{4}{3}$$

Finally, for the inner product we obtain

$$(f, g) = \int_0^1 f(x)g(x)\, dx = \int_0^1 (-2x + 1)(5x^2 - 2x)\, dx$$

$$= \int_0^1 (-10x^3 + 9x^2 - 2x)\, dx$$

Integrating term by term, we obtain

$$(f, g) = \frac{-10}{4}x^4 + 3x^3 - x^2 \Big|_0^1 = \frac{-5}{2} + 3 - 1 = -\frac{1}{2}$$

EXAMPLE 6-5
Are the functions used in the previous example orthogonal?

SOLUTION 6-5
The functions are not orthogonal because $(f, g) \neq 0$.

EXAMPLE 6-6
The functions $\cos\theta$ and $\sin\theta$ belong to $C[0, 2\pi]$. What are their norms? Are they orthonormal?

SOLUTION 6-6
A plot of $\cos\theta$ over the given range is shown in Fig. 6-3.
   The norm is found by calculating

$$\int_0^{2\pi} \cos^2\theta\, d\theta = \int_0^{2\pi} \frac{1 + \cos 2\theta}{2}\, d\theta = \frac{\theta}{2} + \frac{1}{4}\sin 2\theta \Big|_0^{2\pi} = \pi$$

and so the norm is $\sqrt{\pi}$. The sin function is shown in Fig. 6-4.
   We have

$$\int_0^{2\pi} \sin^2\theta\, d\theta = \int_0^{2\pi} \frac{1 - \cos 2\theta}{2}\, d\theta = \frac{\theta}{2} - \frac{1}{4}\sin 2\theta \Big|_0^{2\pi} = \pi$$

and so again, the norm is $\sqrt{\pi}$.
   Recall that we can normalize a vector (in this case a function) by dividing by the norm. Therefore we see that the normalized functions for the vector space

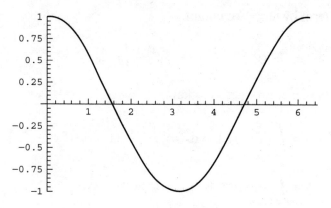

Fig. 6-3. The cos function in the interval defined for $C[0, 2\pi]$.

$C[0, 2\pi]$, found by dividing each function by the norm would be

$$f = \frac{\cos\theta}{\sqrt{\pi}}, \ g = \frac{\sin\theta}{\sqrt{\pi}}$$

If these functions are orthonormal, then

$$\int_0^{2\pi} fg \, d\theta = 0$$

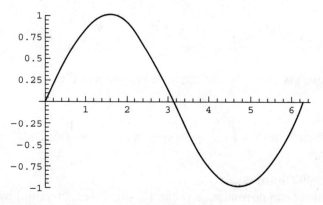

Fig. 6-4. The sin function over the interval defined by the vector space $C[0, 2\pi]$.

In this case we have

$$\frac{1}{\pi} \int_0^{2\pi} \cos\theta \sin\theta \, d\theta$$

Choosing $u = \sin\theta$, we have $du = \cos\theta \, d\theta$ and

$$\frac{1}{\pi} \int_0^{2\pi} \cos\theta \, \sin\theta \, d\theta = \frac{1}{\pi} \frac{\sin^2\theta}{2} \Big|_0^{2\pi} = 0$$

Therefore, the functions $f = \frac{\cos\theta}{\sqrt{\pi}}$, $g = \frac{\sin\theta}{\sqrt{\pi}}$ are orthonormal on $C[0, 2\pi]$.

Properties of the Norm

In Chapter 4 we stated the Cauchy–Schwarz and triangle inequalities. These relations can be used to derive properties of the norm. If a vector space V is an inner product space, then the norm satisfies

- $\|u\| \geq 0$ with $\|u\| = 0$ if and only if $u = 0$
- $\|\alpha u\| = |\alpha| \|u\|$
- $\|u + v\| \leq \|u\| + \|v\|$

You may recall that the last property is the triangle inequality. This is an abstraction of the notion from ordinary geometry that the length of one side of a triangle cannot be longer than the lengths of the other two sides summed together. Using ordinary vectors, we can visualize this by using vector addition (see Fig. 6-5).

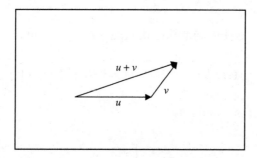

Fig. 6-5. An illustration of the triangle inequality using ordinary vectors.

An Inner Product for Matrix Spaces

The set of $m \times n$ matrices form a vector space which we denote $M_{m,n}$. Suppose that $A, B \in M_{m,n}$ are two $m \times n$ matrices. An inner product exists for this space and is calculated in the following way:

$$(A, B) = \text{tr}\left(B^T A\right)$$

EXAMPLE 6-7
Find the angle between two matrices, $\cos \theta$, where

$$A = \begin{bmatrix} -2 & 1 \\ 4 & 1 \end{bmatrix}, \quad B = \begin{bmatrix} 5 & 0 \\ 1 & 2 \end{bmatrix}$$

SOLUTION 6-7
The inner product is given by

$$(A, B) = \text{tr}\left(B^T A\right)$$

First we compute the transpose of B

$$B^T = \begin{bmatrix} 5 & 1 \\ 0 & 2 \end{bmatrix}$$

And so we have

$$B^T A = \begin{bmatrix} 5 & 1 \\ 0 & 2 \end{bmatrix} \begin{bmatrix} -2 & 1 \\ 4 & 1 \end{bmatrix} = \begin{bmatrix} -6 & 6 \\ 8 & 2 \end{bmatrix}$$

We calculate the trace by summing the diagonal elements

$$(A, B) = \text{tr}\left(B^T A\right) = -6 + 2 = -4$$

The transpose of A is

$$A^T = \begin{bmatrix} -2 & 4 \\ 1 & 1 \end{bmatrix}$$

And so we have

$$A^T A = \begin{bmatrix} -2 & 4 \\ 1 & 1 \end{bmatrix} \begin{bmatrix} -2 & 1 \\ 4 & 1 \end{bmatrix} \quad B = \begin{bmatrix} 20 & 2 \\ 2 & 2 \end{bmatrix}$$

Therefore we find that

$$(A, A) = \operatorname{tr}\left(A^T A\right) = 20 + 2 = 22$$

The norm of A is found by taking the square root of the inner product:

$$\|A\| = \sqrt{(A, A)} = \sqrt{22}$$

For B we have

$$B^T B = \begin{bmatrix} 5 & 1 \\ 0 & 2 \end{bmatrix} \begin{bmatrix} 5 & 0 \\ 1 & 2 \end{bmatrix} = \begin{bmatrix} 26 & 2 \\ 2 & 4 \end{bmatrix}$$

and so

$$(B, B) = \operatorname{tr}\left(B^T B\right) = 26 + 4 = 30$$

The norm of B is

$$\|B\| = \sqrt{(B, B)} = \sqrt{30}$$

Putting these results together, we find

$$\cos \theta = \frac{(A, B)}{\|A\| \, \|B\|} = \frac{-4}{\sqrt{22}\sqrt{30}}$$

The Gram-Schmidt Procedure

An orthonormal basis can be produced from an arbitrary basis by application of the *Gram-Schmidt orthogonalization* process. Let $\{v_1, v_2, \ldots, v_n\}$ be a basis for some inner product space V. The Gram-Schmidt process constructs an orthogonal basis w_i as follows:

$$w_1 = v_1$$

$$w_2 = v_2 - \frac{(w_1, v_2)}{(w_1, w_1)} w_1$$

$$\vdots$$

$$w_n = v_n - \frac{(w_1, v_n)}{(w_1, w_1)} w_1 - \frac{(w_2, v_n)}{(w_2, w_2)} w_n - \cdots - \frac{(w_{n-1}, v_n)}{(w_{n-1}, w_{n-1})} w_{n-1}$$

To form an orthonormal set using this procedure, divide each vector by its norm.

EXAMPLE 6-8

Use the Gram-Schmidt process to construct an orthonormal basis set from

$$v_1 = \begin{bmatrix} 1 \\ 2 \\ -1 \end{bmatrix}, \quad v_2 = \begin{bmatrix} 0 \\ 1 \\ -1 \end{bmatrix}, \quad v_3 = \begin{bmatrix} 3 \\ -7 \\ 1 \end{bmatrix}$$

SOLUTION 6-8

We use a tilde character to denote the unnormalized vectors. The first basis vector is

$$\tilde{w}_1 = v_1$$

Now let's normalize this vector

$$(v_1, v_1) = (1 \quad 2 \quad -1) \begin{pmatrix} 1 \\ 2 \\ -1 \end{pmatrix} = 1 \times 1 + 2 \times 2 + (-1) \times (-1)$$

$$= 1 + 4 + 1 = 6$$

$$\Rightarrow w_1 = \frac{\tilde{w}_1}{\sqrt{(v_1, v_1)}} = \frac{1}{\sqrt{6}} \begin{pmatrix} 1 \\ 2 \\ -1 \end{pmatrix}$$

To find the second vector, first we compute

$$(\tilde{w}_1, v_2) = (1 \quad 2 \quad -1) \begin{pmatrix} 0 \\ 1 \\ -1 \end{pmatrix} = [1^*0 + 2^*1 + (-1)^*(-1)] = 3$$

The first vector is already normalized, so

$$\tilde{w}_2 = v_2 - \frac{(\tilde{w}_1, v_2)}{(\tilde{w}_1, \tilde{w}_1)}\tilde{w}_1 = \begin{pmatrix} 0 \\ 1 \\ -1 \end{pmatrix} - \frac{3}{6}\begin{pmatrix} 1 \\ 2 \\ -1 \end{pmatrix} = \begin{pmatrix} -\frac{1}{2} \\ 0 \\ -\frac{1}{2} \end{pmatrix}$$

Now we normalize

$$\tilde{w}_2, \tilde{w}_2 = \begin{pmatrix} -\frac{1}{2} & 0 & -\frac{1}{2} \end{pmatrix}\begin{pmatrix} -\frac{1}{2} \\ 0 \\ -\frac{1}{2} \end{pmatrix} = \frac{1}{4} + 0 + \frac{1}{4} = \frac{1}{2}$$

and so a second normalized vector is

$$w_2 = \frac{1}{\sqrt{(\tilde{w}_2, \tilde{w}_2)}}\tilde{w}_2 = \sqrt{2}\begin{pmatrix} -\frac{1}{2} \\ 0 \\ -\frac{1}{2} \end{pmatrix} = \begin{pmatrix} -\frac{1}{\sqrt{2}} \\ 0 \\ -\frac{1}{\sqrt{2}} \end{pmatrix}$$

Finally, the third vector is found from

$$\tilde{w}_3 = v_3 - \frac{(\tilde{w}_1, v_3)}{(\tilde{w}_1, \tilde{w}_1)}\tilde{w}_1 - \frac{(\tilde{w}_2, v_3)}{(\tilde{w}_2, \tilde{w}_2)}\tilde{w}_2$$

Now

$$(\tilde{w}_2, v_3) = \begin{pmatrix} -\frac{1}{2} & 0 & -\frac{1}{2} \end{pmatrix}\begin{pmatrix} 3 \\ -7 \\ 1 \end{pmatrix} = -\frac{3}{2} - \frac{1}{2} = -\frac{4}{2} = -2$$

and so

$$\tilde{w}_3 = \begin{pmatrix} 3 \\ -7 \\ 1 \end{pmatrix} + \frac{12}{6}\begin{pmatrix} 1 \\ 2 \\ -1 \end{pmatrix} + \frac{2}{\left(\frac{1}{2}\right)}\begin{pmatrix} -\frac{1}{2} \\ 0 \\ -\frac{1}{2} \end{pmatrix}$$

$$= \begin{pmatrix} 3 \\ -7 \\ 1 \end{pmatrix} + \begin{pmatrix} 2 \\ 4 \\ -2 \end{pmatrix} + \begin{pmatrix} -2 \\ 0 \\ -2 \end{pmatrix} = \begin{pmatrix} 3 \\ -3 \\ -3 \end{pmatrix}$$

Normalizing we find

$$(\tilde{w}_3, \tilde{w}_3) = \begin{pmatrix} 3 & -3 & -3 \end{pmatrix} \begin{pmatrix} 3 \\ -3 \\ -3 \end{pmatrix} = 9 + 9 + 9 = 27$$

and so the last normalized basis vector is

$$w_3 = \frac{1}{\sqrt{(\tilde{w}_3, \tilde{w}_3)}} \tilde{w}_3 = \frac{1}{\sqrt{27}} \begin{pmatrix} 3 \\ -3 \\ -3 \end{pmatrix} = \frac{1}{3\sqrt{3}} \begin{pmatrix} 3 \\ -3 \\ -3 \end{pmatrix} = \frac{1}{\sqrt{3}} \begin{pmatrix} 1 \\ -1 \\ -1 \end{pmatrix}$$

Quiz

1. For the vector space $\mathbb{C}^2$, the inner product is defined by

    $$(u, v) = u_1^* v_1 + u_2^* v_2$$

    Show that $(u, v) = (v, u)^*$ and that the inner product is antilinear in the first argument, but linear in the second argument.

2. Consider the vector space $\mathbb{C}^2$. Let $u, v, w \in \mathbb{C}^2$ and suppose that

    $$(u, v) = 2i$$
    $$(u, w) = 1 + 9i$$

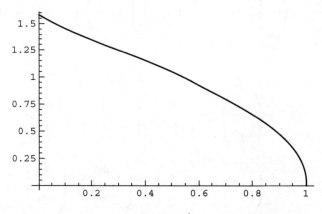

Fig. 6-6. $\cos^{-1}(x)$.

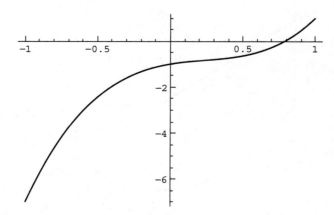

Fig. 6-7. $f(x) = 3x^3 - 2x^2 + x - 1$ shown on $C[-1, 1]$.

Find $(v - 2w, u)$ and $2(3iu, v) - (u, iw)$.
3. Find the inner product of the matrices

$$A = \begin{bmatrix} -1 & 1 \\ 1 & 1 \end{bmatrix}, \quad \text{and} \quad B = \begin{bmatrix} 2 & 3 \\ 4 & 5 \end{bmatrix}$$

4. Consider the vector space of continuous functions $C[0, 1]$. The function is $\cos^{-1}(x)$, which is shown in Fig. 6-6.
    Is it possible to find the norm of $\cos^{-1}(x)$.
5. Find the norm of $f(x) = 3x^3 - 2x^2 + x - 1$ on $C[-1, 1]$ (see Fig. 6-7).

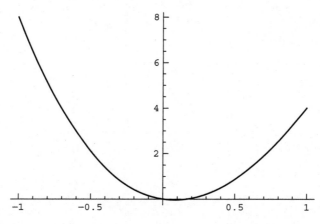

Fig. 6-8. $-x^3 + 6x^2 - x$ shown over the interval defined by $C[-1, 1]$.

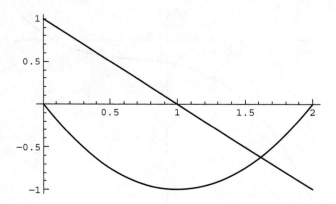

Fig. 6-9. The functions $f(x) = x^2 - 2x$ and $g(x) = -x + 1$ are orthogonal on $C[0, 2]$.

6. Is $f(x) = 3x^3 - 2x^2 + x - 1$ orthogonal to $-x^3 + 6x^2 - x$ on $C[-1, 1]$ (see Fig. 6-8)?

7. Are the columns of

$$A = \begin{bmatrix} 1 & 0 & 4 \\ 2 & 2 & -5 \\ 3 & 5 & 2 \end{bmatrix}$$

orthogonal?

8. Show that the functions $f(x) = x^2 - 2x$ and $g(x) = -x + 1$ are orthogonal on $C[0, 2]$ (see Fig. 6-9). Normalize these functions.

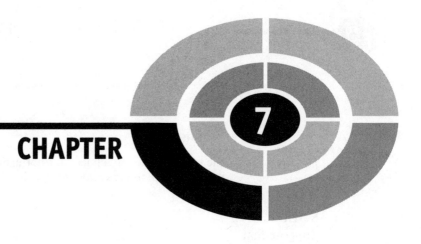

CHAPTER 7

Linear Transformations

Suppose that V and W are two vector spaces. A *linear transformation* T is a function from V to W that has the following properties (see Fig. 7-1):

- $T(v + w) = T(v) + T(w)$
- $T(\alpha v) = \alpha\, T(v)$

EXAMPLE 7-1

Is the function $T : \mathbb{R}^2 \to \mathbb{R}^2$ that swaps vector components

$$T \begin{bmatrix} a \\ b \end{bmatrix} = \begin{bmatrix} b \\ a \end{bmatrix}$$

a linear transformation?

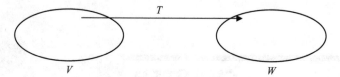

Fig. 7-1. A schematic representation of a linear transformation. T maps vectors from the vector space V to the vector space W in a linear way.

SOLUTION 7-1

Suppose that

$$v = \begin{bmatrix} a \\ b \end{bmatrix} \text{ and } w = \begin{bmatrix} c \\ d \end{bmatrix}$$

are two vectors in $\mathbb{R}^2$.

We check the first property by applying the transformation to the sum of the two vectors:

$$T(v + w) = T\left(\begin{bmatrix} a \\ b \end{bmatrix} + \begin{bmatrix} c \\ d \end{bmatrix} \right) = T\left(\begin{bmatrix} a + c \\ b + d \end{bmatrix} \right) = \begin{bmatrix} b + d \\ a + c \end{bmatrix}$$

Now we first apply the transformation to each of the vectors alone, and then add the results:

$$T(v) + T(w) = T\left(\begin{bmatrix} a \\ b \end{bmatrix} \right) + T\left(\begin{bmatrix} c \\ d \end{bmatrix} \right) = \begin{bmatrix} b \\ a \end{bmatrix} + \begin{bmatrix} d \\ c \end{bmatrix} = \begin{bmatrix} b + d \\ a + c \end{bmatrix}$$

The application of the transformation both ways produces the same vector, indicating that the transformation is linear. We also need to check how the transformation acts on a vector multiplied by a scalar. Let z be some scalar. Then

$$T(zv) = T\left(z \begin{bmatrix} a \\ b \end{bmatrix} \right) = T\left(\begin{bmatrix} za \\ zb \end{bmatrix} \right) = \begin{bmatrix} zb \\ za \end{bmatrix}$$

We also have

$$zT(v) = zT\left(\begin{bmatrix} a \\ b \end{bmatrix} \right) = z \begin{bmatrix} b \\ a \end{bmatrix} = \begin{bmatrix} zb \\ za \end{bmatrix} = T(zv)$$

We conclude that the transformation is linear.

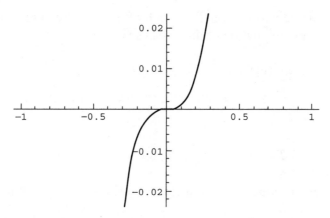

Fig. 7-2. The transformation that takes x to x^3 is not linear.

EXAMPLE 7-2

Is the transformation

$$T(x) = x^3$$

linear?

SOLUTION 7-2

We have

$$T(x) + T(y) = x^3 + y^3$$

but

$$T(x + y) = (x + y)^3 = x^3 + 3x^2y + 3xy^2 + y^3 \neq x^3 + y^3$$

Therefore, the transformation is not linear (see Fig. 7-2).

Matrix Representations

We can represent a linear transformation $T : V \to W$ by a matrix. This is done by finding the matrix representation with respect to a given basis. The matrix is found by applying the transformation to each vector in the basis set. To find the matrix representation, we let $\{v_1, v_2, \ldots, v_n\}$ represent a basis for vector space V and $\{w_1, w_2, \ldots, w_m\}$ represent a basis for vector space W. We then consider the action of the transformation T on each of the basis vectors of V. This will give some linear combination of the w basis vectors:

$$T(v_1) = a_{11}w_1 + a_{21}w_2 + \cdots + a_{m1}w_m$$
$$T(v_2) = a_{12}w_1 + a_{22}w_2 + \cdots + a_{m2}w_m$$
$$\vdots$$
$$T(v_n) = a_{1n}w_1 + a_{2n}w_2 + \cdots + a_{mn}w_m$$

We can arrange the coefficients in these expansions in an $n \times m$ matrix:

$$T = \begin{pmatrix} a_{11} & \cdots & a_{1n} \\ \vdots & \ddots & \vdots \\ a_{m1} & \cdots & a_{mn} \end{pmatrix}$$

This is the matrix representation of the transformation T with respect to the bases from V and W.

To find the matrix representation of a transformation between two vector spaces of dimension n and m over the real field, we apply the following algorithm:

- Apply the transformation to each of the basis vectors of V.
- Construct an augmented matrix of the form

$$\begin{bmatrix} A & | & B \end{bmatrix}$$

The columns of A are the basis vectors of W and the columns of B are the vectors found from the action of T on the basis vectors of V.

Now apply row reduction techniques to transform this matrix into

$$\begin{bmatrix} A & | & B \end{bmatrix} \rightarrow \begin{bmatrix} I & | & T \end{bmatrix}$$

where I is the $m \times m$ identity matrix and T is the matrix representation of the linear transformation. The number of columns in the matrix representation of T is equal to the dimension of vector space V and the number of rows in this matrix is equal to the dimension of the vector space W.

EXAMPLE 7-3
Suppose that we have the linear transformation

$$T(a, b, c) = (a + b, 6a - b + 2c)$$

Find the matrix which represents this transformation with respect to the standard basis of $\mathbb{R}^3$ and the basis

$$w_1 = \begin{bmatrix} 1 \\ 1 \end{bmatrix}, \quad w_2 = \begin{bmatrix} 1 \\ -1 \end{bmatrix}$$

of $W = \mathbb{R}^2$.

SOLUTION 7-3

We call $\mathbb{R}^3$ the vector space V. The standard basis of $\mathbb{R}^3$ is given by

$$e_1 = (1, 0, 0)$$
$$e_2 = (0, 1, 0)$$
$$e_3 = (0, 0, 1)$$

We act T on each of these vectors, obtaining

$$T(1, 0, 0) = (1, 6)$$
$$T(0, 1, 0) = (1, -1)$$
$$T(0, 0, 1) = (0, 2)$$

Now we construct our matrix for reduction. On the left each column is one of the basis vectors of W. On the right, we list the vectors created by action of T on the basis vectors of V (in this case, the standard basis of $\mathbb{R}^3$):

$$\left[\begin{array}{cc|ccc} 1 & 1 & 1 & 1 & 0 \\ 1 & -1 & 6 & -1 & 2 \end{array} \right]$$

Now we perform a reduction on the matrix, with the goal of turning the goal of turning the left-hand side into the identity. Step one is to add the second row to the first and place the result into the first row:

$$R_2 + R_1 \rightarrow R_1$$

which gives

$$\left[\begin{array}{cc|ccc} 2 & 0 & 7 & 0 & 2 \\ 1 & -1 & 6 & -1 & 2 \end{array} \right]$$

Now we multiply the first row by 1/2:

$$\left[\begin{array}{cc} 1 & 0 \\ 1 & -1 \end{array}\middle|\begin{array}{ccc} 7/2 & 0 & 1 \\ 6 & -1 & 2 \end{array}\right]$$

Now multiply the second row by -1:

$$\left[\begin{array}{cc} 1 & 0 \\ -1 & 1 \end{array}\middle|\begin{array}{ccc} 7/2 & 0 & 1 \\ -6 & 1 & -2 \end{array}\right]$$

Now we add the first row to the second, and replace the second row with the result:

$$-R_1 + R_2 \rightarrow R_2$$

$$\left[\begin{array}{cc} 1 & 0 \\ 0 & 1 \end{array}\middle|\begin{array}{ccc} 7/2 & 0 & 1 \\ -5/2 & 1 & -1 \end{array}\right]$$

So we have the identity matrix on the left side, indicating we are done. The matrix representing the transformation T with respect to the bases V and W is

$$T = \begin{bmatrix} 7/2 & 0 & 1 \\ -5/2 & 1 & -1 \end{bmatrix}$$

EXAMPLE 7-4

Let $T : \mathbb{R}^3 \rightarrow \mathbb{R}^2$. Find the matrix representation of

$$T(a, b, c) = (-a + b, \ 2b + 4c)$$

where V is the standard basis of $\mathbb{R}^3$ and W is $[(9, 2), (2, 1)]$

SOLUTION 7-4

We find the action of T on each of the basis vectors of $\mathbb{R}^3$:

$$T(a, b, c) = (-a + b, 2b + 4c)$$
$$\Rightarrow T(1, 0, 0) = (-1, 0)$$
$$T(0, 1, 0) = (1, 2)$$
$$T(0, 0, 1) = (0, 4)$$

Using the basis $[(9, 2), (2, 1)]$, the augmented matrix is

$$\left[\begin{array}{cc|ccc} 9 & 2 & -1 & 1 & 0 \\ 2 & 1 & 0 & 2 & 4 \end{array}\right]$$

Now take $-2R_2 + R_1 \rightarrow R_1$. This gives

$$\left[\begin{array}{cc|ccc} 5 & 0 & -1 & -3 & -8 \\ 2 & 1 & 0 & 2 & 4 \end{array}\right]$$

Now we divide R_1 by 5:

$$\left[\begin{array}{cc|ccc} 1 & 0 & -1/5 & -3/5 & -8/5 \\ 2 & 1 & 0 & 2 & 4 \end{array}\right]$$

We make the substitution $-2R_1 + R_2 \rightarrow R_2$, which gives the identity on the left side:

$$\left[\begin{array}{cc|ccc} 1 & 0 & -1/5 & -3/5 & -8/5 \\ 0 & 1 & 2/5 & 16/5 & 36/5 \end{array}\right]$$

and so, the matrix representation with respect to the two bases given is

$$T = \begin{bmatrix} -1/5 & -3/5 & -8/5 \\ 2/5 & 16/5 & 36/5 \end{bmatrix} = \frac{1}{5}\begin{bmatrix} -1 & -3 & -8 \\ 2 & 16 & 36 \end{bmatrix}$$

EXAMPLE 7-5
Now consider a transformation from $V = \mathbb{R}^2$ to $W = \mathbb{R}^3$ given by

$$T(a, b) = (-a, \ a + b, \ a - b)$$

Find the matrix representation of this transformation where the basis of V is given by

$$[(2, 1), (1, 7)]$$

and the basis of W is the standard basis of $\mathbb{R}^3$.

SOLUTION 7-5
The action on the basis of V is

$$T(a, b) = (-a, \ a+b, \ a-b)$$
$$\Rightarrow T(2, 1) = (-2, 3, 1)$$
$$T(1, 7) = (-1, 8, -6)$$

This time we seek the 3×3 identity matrix. The form that the augmented matrix takes in this case tells us we already have it:

$$\begin{bmatrix} 1 & 0 & 0 & & -2 & -1 \\ 0 & 1 & 0 & | & 3 & 8 \\ 0 & 0 & 1 & & 1 & -6 \end{bmatrix}$$

This is easy to see by writing out the action of T as a linear combination

$$T(2, 1) = (-2, 3, 1) = -2(1, 0, 0) + 3(0, 1, 0) + (0, 0, 1)$$
$$T(1, 7) = (-1, 8, -6) = -(1, 0, 0) + 8(0, 1, 0) - 6(0, 0, 1)$$

The matrix representation is

$$T = \begin{bmatrix} -2 & -1 \\ 3 & 8 \\ 1 & -6 \end{bmatrix}$$

EXAMPLE 7-6
A linear transformation $T : \mathbb{R}^3 \rightarrow \mathbb{R}^2$ has the matrix representation

$$T = \begin{bmatrix} 2 & -1 & 2 \\ 4 & 1 & 5 \end{bmatrix}$$

with respect to the standard basis of $\mathbb{R}^3$ and the basis $[(4, 3), (3, 2)]$. Describe the action of this linear transformation.

SOLUTION 7-6
The first column of the matrix gives us the action of the transformation on $(1, 0, 0)$ and so on, in the form

$$T(v_1) = a_{11}w_1 + a_{21}w_2 + \cdots + a_{m1}w_m$$

Therefore we have

$$T\,(1, 0, 0) = 2\,(4, 3) + 4\,(3, 2) = (8, 6) + (12, 8) = (20, 14)$$

$$T\,(0, 1, 0) = -1\,(4, 3) + 1\,(3, 2) = (-4, -3) + (3, 2) = (-1, -1)$$

$$T\,(0, 0, 1) = 2\,(4, 3) + 5\,(3, 2) = (8, 6) + (15, 10) = (23, 16)$$

We can use this information to find the action on an arbitrary vector. Since we can write

$$(a, b, c) = a\,(1, 0, 0) + b\,(0, 1, 0) + c\,(0, 0, 1)$$

and for a linear transformation L we have

$$L\,(\alpha v) = \alpha L\,(v)$$

where α is a scalar and v is a vector. Therefore the action of the transformation in this problem on an arbitrary vector is

$$T\,(a, b, c) = aT(1, 0, 0) + bT(0, 1, 0) + cT(0, 0, 1)$$
$$= a\,(20, 14) + b\,(-1, -1) + c\,(23, 16)$$
$$= (20a - b + 23c,\ 14a - b + 16c)$$

Linear Transformations in the Same Vector Space

In many physical applications we are concerned with linear transformations or operators that act as

$$T : V \to V$$

Suppose that V is an n-dimensional vector space and a suitable basis for V is $\{v_1, v_2, \ldots, v_n\}$. The matrix representation of the operator with respect to the basis V can be found from taking inner products. The representation of the element at (i, j) is

$$T_{ij} = \left(v_i, Tv_j\right)$$

$$T = \begin{bmatrix} (v_1, Tv_1) & (v_1, Tv_2) & \cdots & (v_1, Tv_n) \\ (v_2, Tv_1) & (v_2, Tv_2) & \cdots & \vdots \\ \vdots & \vdots & \ddots & \vdots \\ (v_n, Tv_1) & (v_n, Tv_2) & \cdots & (v_n, Tv_n) \end{bmatrix}$$

EXAMPLE 7-7

Consider a three-dimensional vector space with an orthonormal basis $\{u_1, u_2, u_3\}$. An operator A acts on this basis in the following way:

$$Au_1 = u_2 + 4u_3$$

$$Au_2 = 2u_1$$

$$Au_3 = u_1 - u_3$$

Find the matrix representation of this operator with respect to this basis.

SOLUTION 7-7

The basis is orthonormal, and so we have

$$\left(u_i, u_j\right) = \delta_{ij}$$

The matrix representation is

$$A = \begin{bmatrix} (u_1, Au_1) & (u_1, Au_2) & (u_1, Au_3) \\ (u_2, Au_1) & (u_2, Au_2) & (u_2, Au_3) \\ (u_3, Au_1) & (u_3, Au_2) & (u_3, Au_3) \end{bmatrix}$$

Using the action of the operator on the states, we have

$$A = \begin{bmatrix} (u_1, u_2) + 4(u_1, u_3) & 2(u_1, u_1) & (u_1, u_1) - (u_1, u_3) \\ (u_2, u_2) + 4(u_2, u_3) & 2(u_2, u_1) & (u_2, u_1) - (u_2, u_3) \\ (u_3, u_2) + 4(u_3, u_3) & 2(u_3, u_1) & (u_3, u_1) - (u_3, u_3) \end{bmatrix}$$

$$= \begin{bmatrix} 0 & 2 & 1 \\ 1 & 0 & 0 \\ 4 & 0 & -1 \end{bmatrix}$$

EXAMPLE 7-8

Now we consider a two-dimensional complex vector space. A basis for the space is

$$v_1 = \begin{bmatrix} 1 \\ 0 \end{bmatrix}, \qquad v_2 = \begin{bmatrix} 0 \\ 1 \end{bmatrix}$$

Hadamard operator H acts on the basis vectors in the following way:

$$Hv_1 = \frac{v_1 + v_2}{\sqrt{2}}, \qquad Hv_2 = \frac{v_1 - v_2}{\sqrt{2}}$$

Find the matrix representation of H in this basis, which is orthornormal.

SOLUTION 7-8

The matrix representation is

$$H \doteq \begin{bmatrix} (v_1, Hv_1) & (v_1, Hv_2) \\ (v_2, Hv_1) & (v_2, Hv_2) \end{bmatrix}$$

Using the action of H on the basis states, we obtain

$$H \doteq \begin{bmatrix} \left(v_1, \frac{v_1+v_2}{\sqrt{2}}\right) & \left(v_1, \frac{v_1-v_2}{\sqrt{2}}\right) \\ \left(v_2, \frac{v_1+v_2}{\sqrt{2}}\right) & \left(v_2, \frac{v_1-v_2}{\sqrt{2}}\right) \end{bmatrix}$$

$$= \frac{1}{\sqrt{2}} \begin{bmatrix} (v_1, v_1) + (v_1, v_2) & (v_1, v_1) - (v_1, v_2) \\ (v_2, v_1) + (v_2, v_2) & (v_2, v_1) - (v_2, v_2) \end{bmatrix}$$

$$= \frac{1}{\sqrt{2}} \begin{bmatrix} 1 & 1 \\ 1 & -1 \end{bmatrix}$$

EXAMPLE 7-9

A linear transformation $L : \mathbb{R}^3 \to \mathbb{R}^3$ acts as

$$L(a, b, c) = (a + b, 3a - 2c, 2a + 4c)$$

Find the matrix representation with respect to the standard basis.

SOLUTION 7-9

Using

$$L\,(a, b, c) = (a + b, 3a - 2c, 2a + 4c)$$

We find the action of this transformation on the basis vectors to be

$$L\,(1, 0, 0) = (1, 3, 2)$$
$$L\,(0, 1, 0) = (1, 0, 0)$$
$$L\,(0, 0, 1) = (0, -2, 4)$$

The matrix representation is

$$L \doteq \begin{bmatrix} (e_1, Le_1) & (e_1, Le_2) & (e_1, Le_3) \\ (e_2, Le_1) & (e_2, Le_2) & (e_2, Le_3) \\ (e_3, Le_1) & (e_3, Le_2) & (e_3, Le_3) \end{bmatrix}$$

We find the matrix elements by taking the inner products with the vectors that result from the action of L on the standard basis we found above. The first element is

$$(e_1, Le_1) = \begin{bmatrix} 1 & 0 & 0 \end{bmatrix} \begin{bmatrix} 1 \\ 3 \\ 2 \end{bmatrix} = (1)(1) + (0)(3) + (0)(2) = 1$$

Moving down the first column, the next element is

$$(e_2, Le_1) = \begin{bmatrix} 0 & 1 & 0 \end{bmatrix} \begin{bmatrix} 1 \\ 3 \\ 2 \end{bmatrix} = (0)(1) + (1)(3) + (0)(2) = 3$$

The last element in the column is

$$(e_3, Le_1) = \begin{bmatrix} 0 & 0 & 1 \end{bmatrix} \begin{bmatrix} 1 \\ 3 \\ 2 \end{bmatrix} = (0)(1) + (0)(3) + (1)(2) = 2$$

Now we compute the elements in the second column. The top element is

$$(e_1, Le_2) = \begin{bmatrix} 1 & 0 & 0 \end{bmatrix} \begin{bmatrix} 1 \\ 0 \\ 0 \end{bmatrix} = (1)(1) + (0)(0) + (0)(0) = 1$$

The next element is

$$(e_2, Le_2) = \begin{bmatrix} 0 & 1 & 0 \end{bmatrix} \begin{bmatrix} 1 \\ 0 \\ 0 \end{bmatrix} = (0)(1) + (1)(0) + (0)(0) = 0$$

and the last element in the second column is

$$(e_3, Le_2) = \begin{bmatrix} 0 & 0 & 1 \end{bmatrix} \begin{bmatrix} 1 \\ 0 \\ 0 \end{bmatrix} = (0)(1) + (0)(0) + (1)(0) = 0$$

The first element of the third column is

$$(e_1, Le_3) = \begin{bmatrix} 1 & 0 & 0 \end{bmatrix} \begin{bmatrix} 0 \\ -2 \\ 4 \end{bmatrix} = (1)(0) + (0)(-2) + (0)(4) = 0$$

The middle element of the third column is

$$(e_2, Le_3) = \begin{bmatrix} 0 & 1 & 0 \end{bmatrix} \begin{bmatrix} 0 \\ -2 \\ 4 \end{bmatrix} = (0)(0) + (1)(-2) + (0)(4) = -2$$

and the last element in the matrix is

$$(e_3, Le_3) = \begin{bmatrix} 0 & 0 & 1 \end{bmatrix} \begin{bmatrix} 0 \\ -2 \\ 4 \end{bmatrix} = (0)(0) + (0)(-2) + (1)(4) = 4$$

Putting all of these results together, we obtain the matrix representation with respect to the standard basis:

$$L \doteq \begin{bmatrix} (e_1, Le_1) & (e_1, Le_2) & (e_1, Le_3) \\ (e_2, Le_1) & (e_2, Le_2) & (e_2, Le_3) \\ (e_3, Le_1) & (e_3, Le_2) & (e_3, Le_3) \end{bmatrix} = \begin{bmatrix} 1 & 1 & 0 \\ 3 & 0 & -2 \\ 2 & 0 & 4 \end{bmatrix}$$

EXAMPLE 7-10
Is the transformation

$$T(a, b, c) = (a + 2, \; b - c, \; 5c)$$

linear?

SOLUTION 7-10
We have

$$T(0, 0, 0) = (2, 0, 0) \neq (0, 0, 0)$$

Therefore the transformation is not linear.

EXAMPLE 7-11
Is the transformation

$$T(a, b, c) = (4a - 2b, bc, c)$$

linear?

SOLUTION 7-11
Consider two vectors $u = (a, b, c)$ and $v = (x, y, z)$.
  The sum of the transformations of these vectors is

$$T(u) + T(v) = (4a - 2b, bc, c) + (4x - 2y, yz, z)$$
$$= [4(a + x) - 2(b + y), bc + yz, c + z]$$

But we have

$$T(u + v) = T(a + x, b + y, c + z)$$
$$= [4(a + x) - 2(b + y), (b + y)(c + z), c + z]$$
$$= [4(a + x) - 2(b + y), bc + bz + cy + yz, c + z]$$
$$\neq T(u) + T(v)$$

Therefore this transformation cannot be linear.

More Properties of Linear Transformations

If T and S are two linear transformations and v is some vector, then

$$(T + S)(v) = T(v) + S(v)$$

The product of two linear operators is defined by

$$(TS)(v) = T[S(v)]$$

EXAMPLE 7-12

Let

$$T(x, y, z) = (3x, 2y - z)$$
$$S(x, y, z) = (x, -z)$$

be two linear transformations from $\mathbb{R}^3 \to \mathbb{R}^2$. Find $T + S$, $2T$, and $T - 4S$.

SOLUTION 7-12

Using linearity, we have

$$(T + S)(x, y, z) = T(x, y, z) + S(x, y, z) = (3x, 2y - z) + (x, -z)$$
$$= (4x, 2y - 2z)$$

We also use linearity to find the second transformation

$$2T(x, y, z) = 2(3x, 2y - z) = (6x, 4y - 2z)$$

For the last transformation, we have

$$(T - 4S)(x, y, z) = T(x, y, z) - 4S(x, y, z) = (3x, 2y - z) - 4(x, -z)$$
$$= (-x, 2y + 3z)$$

EXAMPLE 7-13

Consider a linear transformation on polynomials that acts from $P_2 \to P_1$ (i.e., from second-order to first-order polynomials) in the following way:

$$L(ax^2 + bx + c) = (2a - c)t + (a + b + c)$$

Find the matrix that represents this transformation with respect to the bases

$$\{2x^2 + x, 3x, x + 1\} \text{ and } \{2x + 1, x\}$$

for P_2 and P_1, respectively.

SOLUTION 7-13
We make the identification of the polynomial

$$ax^2 + bx + c$$

with the vector (a, b, c). This will allows us to map the problem into a transformation $\mathbb{R}^3 \rightarrow \mathbb{R}^2$. Therefore the basis can be identified as

$$\{2x^2 + x, 3x, x + 1\}$$
$$2x^2 + x \rightarrow (2, 1, 0)$$
$$3x \rightarrow (0, 3, 0)$$
$$x + 1 \rightarrow (0, 1, 1)$$

and we identify $\{2x + 1, x\}$ with

$$2x + 1 \rightarrow (2, 1)$$
$$x \rightarrow (1, 0)$$

The transformation can be restated as

$$L\left(ax^2 + bx + c\right) = (2a - c)t + (a + b + c)$$
$$\rightarrow L(a, b, c) = (2a - c, a + b + c)$$

Now we solve the problem in the same way we solved the others. We first consider the action of the transformation on each of the vectors of $\mathbb{R}^3$:

$$L(2, 1, 0) = (4, 3)$$
$$L(0, 3, 0) = (0, 3)$$
$$L(0, 1, 1) = (-1, 2)$$

Our mapping to a basis for $\mathbb{R}^2$ gave us

$$2x + 1 \rightarrow (2, 1)$$
$$x \rightarrow (1, 0)$$

So the augmented matrix is

$$\left[\begin{array}{cc|ccc} 2 & 1 & 4 & 0 & -1 \\ 1 & 0 & 3 & 3 & 2 \end{array}\right]$$

First we swap rows 1 and 2:

$$\left[\begin{array}{cc|ccc} 1 & 0 & 3 & 3 & 2 \\ 2 & 1 & 4 & 0 & -1 \end{array}\right]$$

Now we make the substitution $-2R_1 + R_2 \rightarrow R_2$, which gives

$$\left[\begin{array}{cc|ccc} 1 & 0 & 3 & 3 & 2 \\ 0 & 1 & -2 & -6 & -5 \end{array}\right]$$

We have the identity matrix in the left block. Therefore we are done and the matrix representation of the transformation with respect to these two bases is

$$\begin{bmatrix} 3 & 3 & 2 \\ -2 & -6 & -5 \end{bmatrix}$$

Quiz

1. Are the following transformations linear?
    (a) $F(x, y, z) = (2x + z, 4y, 8y - 4z)$
    (b) $G(x, y, z) = (2x + 2y, z)$
    (c) $H(x, y, z) = (xy, z)$
    (d) $T(x, y, z) = (2 + x, y - z + xy)$

2. Find the matrix that represents the transformation

$$T(x, y, z) = (-3x + z, 2y)$$

from $\mathbb{R}^3 \rightarrow \mathbb{R}^2$ with respect to the bases $\{(1, 0, 0), (0, 1, 0), (0, 0, 1)\}$ and $\{(1, 1), (1, -1)\}$.

3. Find the matrix that represents the transformation

$$T(x, y, z) = (4x + y + z, y - z)$$

from $\mathbb{R}^3 \to \mathbb{R}^2$ with respect to the bases $\{(1, 1, 0), (-1, 3, 5),$ $(2, -5, 1)\}$ and $\{(1, 1), (1, -1)\}$.

4. Suppose an operator acts

$$Zv_1 = v_1$$
$$Zv_2 = -v_2$$

where

$$v_1 = \begin{bmatrix} 1 \\ 0 \end{bmatrix}, \qquad v_2 = \begin{bmatrix} 0 \\ 1 \end{bmatrix}$$

Find the matrix representation of Z with respect to this basis.

5. Describe the transformation from $\mathbb{R}^3 \to \mathbb{R}^2$ that has the matrix representation

$$T = \begin{bmatrix} 1 & 2 & 5 \\ 4 & -1 & 2 \end{bmatrix}$$

with respect to the standard basis of $\mathbb{R}^3$ and with respect to $\{(1, 1),$ $(1, -1)\}$ for $\mathbb{R}^2$.

6. A linear transformation that acts $\mathbb{R}^3 \to \mathbb{R}^3$ is

$$T(x, y, z) = (2x + y + z, y - z, 4x - 2y + 8z)$$

Find the matrix representation of this transformation with respect to the standard basis.

7. A transformation from $P_2 \to P_1$ acts as

$$T\left(ax^2 + bx + c\right) = (2a + b)x + (b - c)$$

Find the matrix representation of T with respect to the basis

$$\left\{-x^2 + 3x + 5, x^2 - 7x + 1, x^2 + x\right\}$$

for P_2 and with respect to $\{2x + 1, x - 1\}$ for P_1.

8. Let $F(x, y, z) = (2x + y, z)$ and $G(x, y, z) = (4x + z, y - 4z)$.
    Describe
    (a) $F + G$
    (b) $3F$
    (c) $2G$
    (d) $2F - G$

9. An operator acts on a two-dimensional orthonormal basis of $\mathbb{C}^2$ in the
    following way:

    $$Av_1 = 2v_1 - iv_2$$

    $$Av_2 = 4v_2$$

    Find the matrix representation of A with respect to this basis.

10. Suppose a transformation from $\mathbb{R}^2 \rightarrow \mathbb{R}^3$ is represented by

    $$T = \begin{bmatrix} 1 & 0 \\ 2 & 4 \\ 7 & 3 \end{bmatrix}$$

    with respect to the basis $\{(2, 1), (1, 5)\}$ and the standard basis of $\mathbb{R}^3$.
    What are $T(1, 4)$ and $T(3, 5)$?

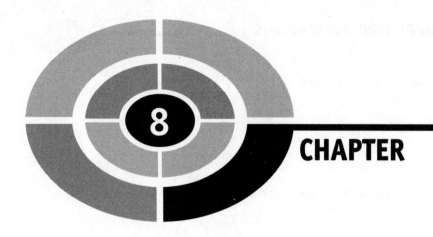

8 CHAPTER

The Eigenvalue Problem

Let A be an $n \times n$ matrix, v an $n \times 1$ column vector, and λ a scalar. If

$$Av = \lambda v$$

we say that v is an *eigenvector* of A and that λ is an *eigenvalue* of A. We now investigate the procedure used to find the eigenvalues of a given matrix.

The Characteristic Polynomial

The characteristic polynomial of a square $n \times n$ matrix A is

$$\Delta(\lambda) = \lambda^n - S_1\lambda^{n-1} + S_2\lambda^{n-2} + \cdots + (-1)^n S_n$$

where S_i are the sum of the principle minors of order i. Less formally, the characteristic polynomial of A is given by

$$\det |A - \lambda I|$$

where λ is an unknown variable and I is the $n \times n$ identity matrix.

The Cayley-Hamilton Theorem

A linear operator A is a zero of its characteristic polynomial.

   In practical calculations, we set the characteristic polynomial equal to zero, giving the *characteristic equation*

$$\det |A - \lambda I| = 0$$

The zeros of the characteristic polynomial, which are the solutions to this equation, are the eigenvalues of the matrix A.

EXAMPLE 8-1
Find the eigenvalues of the matrix

$$A = \begin{bmatrix} 5 & 2 \\ 9 & 2 \end{bmatrix}$$

SOLUTION 8-1
To find the eigenvalues, we solve

$$\det |A - \lambda I| = 0$$

where I is the 2×2 identity matrix

$$I = \begin{bmatrix} 1 & 0 \\ 0 & 1 \end{bmatrix}$$

$$\Rightarrow \lambda I = \begin{bmatrix} \lambda & 0 \\ 0 & \lambda \end{bmatrix}$$

The characteristic polynomial in this case is

$$\det |A - \lambda I| = \det \left| \begin{bmatrix} 5 & 2 \\ 9 & 2 \end{bmatrix} - \begin{bmatrix} \lambda & 0 \\ 0 & \lambda \end{bmatrix} \right| = \det \begin{vmatrix} 5 - \lambda & 2 \\ 9 & 2 - \lambda \end{vmatrix}$$

$$= (5 - \lambda)(2 - \lambda) - 18$$

$$= 10 - 7\lambda + \lambda^2 - 18 = \lambda^2 - 7\lambda - 8$$

Setting this equal to zero gives the characteristic equation

$$\lambda^2 - 7\lambda - 8 = 0$$

This equation factors easily to (check)

$$(\lambda - 8)(\lambda + 1) = 0$$

The solutions of this equation are the two eigenvalues of the matrix:

$$\lambda_1 = 8$$
$$\lambda_2 = -1$$

Notice that for a 2×2 matrix, we found two eigenvalues. This is because a 2×2 matrix leads to a second-order characteristic polynomial. This is true in general; an $n \times n$ matrix will lead to an nth-order characteristic polynomial with n (not necessarily distinct) solutions.

EXAMPLE 8-2
Show that

$$A = \begin{bmatrix} 5 & 2 \\ 9 & 2 \end{bmatrix}$$

satisfies the Cayley-Hamilton Theorem.

SOLUTION 8-2
The characteristic equation for this matrix is

$$\lambda^2 - 7\lambda - 8 = 0$$

The Cayley-Hamilton theorem tells us that

$$A^2 - 7A - 8I = 0$$

Notice that when a constant appears alone in the equation, we insert the identity matrix. First we calculate the square of the matrix A

$$A^2 = \begin{bmatrix} 5 & 2 \\ 9 & 2 \end{bmatrix}\begin{bmatrix} 5 & 2 \\ 9 & 2 \end{bmatrix} = \begin{bmatrix} (5)(5)+(2)(9) & (5)(2)+(2)(2) \\ (9)(5)+(2)(9) & (9)(2)+(2)(2) \end{bmatrix}$$

$$= \begin{bmatrix} 25+18 & 10+4 \\ 45+18 & 18+4 \end{bmatrix} = \begin{bmatrix} 43 & 14 \\ 63 & 22 \end{bmatrix}$$

The second term is

$$7A = (7)\begin{bmatrix} 5 & 2 \\ 9 & 2 \end{bmatrix} = \begin{bmatrix} (7)5 & (7)2 \\ (7)9 & (7)2 \end{bmatrix} = \begin{bmatrix} 35 & 14 \\ 63 & 14 \end{bmatrix}$$

and lastly we have

$$8I = 8\begin{bmatrix} 1 & 0 \\ 0 & 1 \end{bmatrix} = \begin{bmatrix} 8 & 0 \\ 0 & 8 \end{bmatrix}$$

Putting these together, we obtain

$$A^2 - 7A - 8I = \begin{bmatrix} 43 & 14 \\ 63 & 22 \end{bmatrix} - \begin{bmatrix} 35 & 14 \\ 63 & 14 \end{bmatrix} - \begin{bmatrix} 8 & 0 \\ 0 & 8 \end{bmatrix}$$

$$= \begin{bmatrix} 43-35-8 & 14-14 \\ 63-63 & 22-14-8 \end{bmatrix} = 0$$

This verifies the Cayley-Hamilton theorem for this matrix.

EXAMPLE 8-3
Find the eigenvalues of

$$B = \begin{bmatrix} 2 & 1 & 0 \\ 1 & 4 & 0 \\ 2 & 5 & 2 \end{bmatrix}$$

SOLUTION 8-3
The characteristic polynomial is given by

$$\det |B - \lambda I|$$

where I is the 3×3 identity matrix and

$$I = \begin{bmatrix} 1 & 0 & 0 \\ 0 & 1 & 0 \\ 0 & 0 & 1 \end{bmatrix}$$

$$\Rightarrow \lambda I = \begin{bmatrix} \lambda & 0 & 0 \\ 0 & \lambda & 0 \\ 0 & 0 & \lambda \end{bmatrix}$$

Therefore the characteristic polynomial is

$$\det |B - \lambda I| = \det \left| \begin{bmatrix} 2 & 1 & 0 \\ 1 & 4 & 0 \\ 2 & 5 & 2 \end{bmatrix} - \begin{bmatrix} \lambda & 0 & 0 \\ 0 & \lambda & 0 \\ 0 & 0 & \lambda \end{bmatrix} \right|$$

$$= \det \left| \begin{bmatrix} 2 - \lambda & 1 & 0 \\ 1 & 4 - \lambda & 0 \\ 2 & 5 & 2 - \lambda \end{bmatrix} \right|$$

$$= (2 - \lambda) \det \begin{vmatrix} 4 - \lambda & 0 \\ 5 & 2 - \lambda \end{vmatrix} - \det \begin{vmatrix} 1 & 0 \\ 2 & 2 - \lambda \end{vmatrix}$$

$$= (2 - \lambda) [(4 - \lambda)(2 - \lambda)] - (2 - \lambda)$$

$$= (2 - \lambda) [(4 - \lambda)(2 - \lambda) - 1]$$

$$= (2 - \lambda) [8 - 6\lambda + \lambda^2 - 1]$$

$$= (2 - \lambda) [\lambda^2 - 6\lambda + 7]$$

The characteristic equation is obtained by setting this equal to zero

$$(2 - \lambda) [\lambda^2 - 6\lambda + 7] = 0$$

Each term in the product must be zero. We immediately obtain the first eigenvalue by setting the term on the left equal to zero

$$2 - \lambda = 0$$

$$\Rightarrow \lambda_1 = 2$$

To find the other eigenvalues, we solve

$$\lambda^2 - 6\lambda + 7 = 0$$

We find the solutions by recalling the quadratic formula. If the equation is

$$a\lambda^2 + b\lambda + c = 0$$

then

$$\lambda_{2,3} = \frac{-b \pm \sqrt{b^2 - 4ac}}{2a}$$

In this case, the quadratic formula gives the solutions

$$\lambda_2 = \frac{6 + \sqrt{36 - 4\,(7)}}{2} = \frac{6 + \sqrt{8}}{2} = \frac{6 + \sqrt{(4)\,(2)}}{2} = \frac{6 + 2\sqrt{2}}{2} = 3 + \sqrt{2}$$

$$\lambda_3 = \frac{6 - \sqrt{36 - 4\,(7)}}{2} = \frac{6 - \sqrt{8}}{2} = 3 - \sqrt{2}$$

Notice that since B is a 3×3 matrix, it has three eigenvalues.

Finding Eigenvectors

The second step in solving the eigenvalue problem is to find the eigenvectors that correspond to each eigenvalue found in the solution of the characteristic equation. It is best to illustrate the procedure with an example.

EXAMPLE 8-4
Find the eigenvectors of

$$A = \begin{bmatrix} 5 & 2 \\ 9 & 2 \end{bmatrix}$$

SOLUTION 8-4
We have already determined that the eigenvalues of this matrix are

$$\lambda_1 = 8$$
$$\lambda_2 = -1$$

We consider each eigenvalue in turn. An eigenvector of a 2×2 matrix is going to be a column vector with 2 components. If we call these two unknowns x and

y, then we can write the vector as

$$v = \begin{bmatrix} x \\ y \end{bmatrix}$$

For the first eigenvalue, the eigenvector equation is

$$Av = \lambda_1 v$$

Specifically, we have the matrix equation

$$Av = \begin{bmatrix} 5 & 2 \\ 9 & 2 \end{bmatrix} \begin{bmatrix} x \\ y \end{bmatrix} = 8 \begin{bmatrix} x \\ y \end{bmatrix}$$

We perform the matrix multiplication on the left side first. Remember, the multiplication AB where A is an $m \times n$ matrix and B is an $n \times p$ matrix results in an $m \times p$ matrix. Therefore if we multiply the 2×2 matrix A with the 2×1 column vector v, we obtain another 2×1 column vector. The components are

$$\begin{bmatrix} 5 & 2 \\ 9 & 2 \end{bmatrix} \begin{bmatrix} x \\ y \end{bmatrix} = \begin{bmatrix} 5x + 2y \\ 9x + 2y \end{bmatrix}$$

Setting this equal to the right side of the eigenvector equation, we have

$$\begin{bmatrix} 5x + 2y \\ 9x + 2y \end{bmatrix} = 8 \begin{bmatrix} x \\ y \end{bmatrix} = \begin{bmatrix} 8x \\ 8y \end{bmatrix}$$

This means we have two equations:

$$5x + 2y = 8x$$

$$9x + 2y = 8y$$

We use the first equation to write y in terms of x

$$5x + 2y = 8x$$

$$\Rightarrow y = \frac{3}{2}x$$

The system has only one free variable. So we can choose $x = 2$, then $y = 3$. Then the eigenvector corresponding to $\lambda_1 = 8$ is

$$v_1 = \begin{bmatrix} 2 \\ 3 \end{bmatrix}$$

We check this result

$$Av_1 = \begin{bmatrix} 5 & 2 \\ 9 & 2 \end{bmatrix} \begin{bmatrix} 2 \\ 3 \end{bmatrix} = \begin{bmatrix} (5)(2) + (2)(3) \\ (9)(2) + (2)(3) \end{bmatrix} = \begin{bmatrix} 16 \\ 24 \end{bmatrix} = 8 \begin{bmatrix} 2 \\ 3 \end{bmatrix} = \lambda_1 v_1$$

Now we consider the second eigenvalue, $\lambda_2 = -1$. The eigenvector equation is

$$Av_2 = \lambda_2 v_2$$

$$\Rightarrow \begin{bmatrix} 5 & 2 \\ 9 & 2 \end{bmatrix} \begin{bmatrix} x \\ y \end{bmatrix} = \begin{bmatrix} 5x + 2y \\ 9x + 2y \end{bmatrix} = - \begin{bmatrix} x \\ y \end{bmatrix}$$

This gives the two equations

$$5x + 2y = -x$$
$$9x + 2y = -y$$

We add these equations together to find

$$14x + 4y = -x - y$$
$$\Rightarrow y = -3x$$

Choosing $x = 1$ gives $y = -3$ and we find an eigenvector

$$v_2 = \begin{bmatrix} 1 \\ -3 \end{bmatrix}$$

Summarizing, we have found the eigenvectors of the matrix A to be

$$v_1 = \begin{bmatrix} 2 \\ 3 \end{bmatrix} \quad \text{with eigenvalue } \lambda_1 = 8$$

$$v_2 = \begin{bmatrix} 1 \\ -3 \end{bmatrix} \quad \text{with eigenvalue } \lambda_2 = -1$$

Normalization

In many applications, such as quantum theory, it is necessary to *normalize* the eigenvectors. This means that the norm of the eigenvector is 1:

$$1 = v^\dagger v$$

The process of finding a solution to this equation is called normalization. When an application forces us to apply normalization, this puts an additional constraint on the components of the vectors. In the previous example we were free to choose the value of x. However, if we required that the eigenvectors were normalized, then the equation $1 = v^\dagger v$ would dictate the value of x.

EXAMPLE 8-5
Find the normalized eigenvectors of the matrix

$$A = \begin{bmatrix} 2 & 0 & 1 \\ 1 & -1 & 0 \\ 3 & 0 & 4 \end{bmatrix}$$

SOLUTION 8-5
The first step is to find the eigenvalues of the matrix. We begin by deriving the characteristic polynomial. We have

$$\det |A - \lambda I| = \det \left| \begin{bmatrix} 2 & 0 & 1 \\ 1 & -1 & 0 \\ 3 & 0 & 4 \end{bmatrix} - \lambda \begin{bmatrix} 1 & 0 & 0 \\ 0 & 1 & 0 \\ 0 & 0 & 1 \end{bmatrix} \right|$$

$$= \det \left| \begin{bmatrix} 2 & 0 & 1 \\ 1 & -1 & 0 \\ 3 & 0 & 4 \end{bmatrix} - \begin{bmatrix} \lambda & 0 & 0 \\ 0 & \lambda & 0 \\ 0 & 0 & \lambda \end{bmatrix} \right|$$

This gives

$$\det \begin{vmatrix} 2-\lambda & 0 & 1 \\ 1 & -1-\lambda & 0 \\ 3 & 0 & 4-\lambda \end{vmatrix} = (2-\lambda) \det \begin{vmatrix} -1-\lambda & 0 \\ 0 & 4-\lambda \end{vmatrix} + \det \begin{vmatrix} 1 & -1-\lambda \\ 3 & 0 \end{vmatrix}$$

$$= (2-\lambda)[(-1-\lambda)(4-\lambda)] - (3)(-1-\lambda)$$

$$= (-1-\lambda)[(2-\lambda)(4-\lambda) - 3]$$

$$= (-1-\lambda)[\lambda^2 - 6\lambda + 5]$$

We set this equal to zero to obtain the characteristic equation

$$(-1 - \lambda)\left[\lambda^2 - 6\lambda + 5\right] = 0$$
$$\Rightarrow 1 + \lambda = 0 \text{ or } \lambda = -1$$
$$\lambda^2 - 6\lambda + 5 = (\lambda - 5)(\lambda - 1) = 0 \text{ or } \lambda = 5, \ \lambda = 1$$

So we have three distinct eigenvalues $\{-1, 5, 1\}$. We compute each eigenvector in turn. Starting with $\lambda_1 = -1$ the eigenvector equation is

$$A v_1 = -v_1$$

We set the eigenvector equal to

$$v_1 = \begin{bmatrix} x \\ y \\ z \end{bmatrix}$$

where x, y, z are three unknowns. Applying the matrix A gives us

$$A v_1 = \begin{bmatrix} 2 & 0 & 1 \\ 1 & -1 & 0 \\ 3 & 0 & 4 \end{bmatrix} \begin{bmatrix} x \\ y \\ z \end{bmatrix} = \begin{bmatrix} 2x + z \\ x - y \\ 3x + 4z \end{bmatrix}$$

Setting this equal to $-v_1$ gives three equations

$$2x + z = -x$$
$$x - y = -y$$
$$3x + 4z = -z$$

The second equation immediately tells us that $x = 0$. (We add y to both sides):

$$x - y = -y, \ \Rightarrow x = 0$$

Using $x = 0$ in the third equation, we have

$$3x + 4z = -z, \quad x = 0$$
$$\Rightarrow \quad 4z = -z$$

Which can be true only if $z = 0$ as well. This leaves us with

$$v_1 = \begin{bmatrix} 0 \\ y \\ 0 \end{bmatrix}$$

Under general conditions, y can be any value we choose. However, in this case we require that the eigenvectors be normalized. Therefore we must have

$$v_1^\dagger v_1 = 1$$

We compute this product. We consider the most general case; therefore, we allow y to be a complex number. The Hermitian conjugate of the eigenvector is

$$v_1^\dagger = \begin{bmatrix} 0 & y^* & 0 \end{bmatrix}$$

Therefore the product is

$$v_1^\dagger v_1 = \begin{bmatrix} 0 & y^* & 0 \end{bmatrix} \begin{bmatrix} 0 \\ y \\ 0 \end{bmatrix} = 0 + y^*y + 0 = |y|^2$$

Setting this equal to unity, we find that

$$|y|^2 = 1$$
$$\Rightarrow \quad y = 1$$

up to an undetermined phase, which we are free to discard. Therefore the first eigenvector is

$$v_1 = \begin{bmatrix} 0 \\ 1 \\ 0 \end{bmatrix}$$

Now we consider the second eigenvalue, $\lambda_2 = 5$. This gives

$$Av_2 = \begin{bmatrix} 2 & 0 & 1 \\ 1 & -1 & 0 \\ 3 & 0 & 4 \end{bmatrix} \begin{bmatrix} x \\ y \\ z \end{bmatrix} = \begin{bmatrix} 2x + z \\ x - y \\ 3x + 4z \end{bmatrix} = 5v_2 = \begin{bmatrix} 5x \\ 5y \\ 5z \end{bmatrix}$$

$$\Rightarrow 2x + z = 5x$$
$$x - y = 5y$$
$$3x + 4z = 5z$$

From these equations we obtain

$$z = 3x$$
$$x = 6y$$

and so we can write the eigenvector as

$$v_2 = \begin{bmatrix} x \\ y \\ z \end{bmatrix} = \begin{bmatrix} 6y \\ y \\ 3x \end{bmatrix} = \begin{bmatrix} 6y \\ y \\ 18y \end{bmatrix}$$

The Hermitian conjugate of the vector is

$$v_2^\dagger = \begin{bmatrix} 6y^* & y^* & 18y^* \end{bmatrix}$$

Normalizing we find

$$v_2^\dagger v_2 = \begin{bmatrix} 6y^* & y^* & 18y^* \end{bmatrix} \begin{bmatrix} 6y \\ y \\ 18y \end{bmatrix} = 36\,|y|^2 + |y|^2 + 324\,|y|^2 = 361\,|y|^2$$

$$v_2^\dagger v_2 = 1$$
$$\Rightarrow |y|^2 = \frac{1}{361} \text{ or } y = \frac{1}{\sqrt{361}}$$

This gives

$$v_2 = \begin{bmatrix} 6y \\ y \\ 30y \end{bmatrix} = \frac{1}{\sqrt{361}} \begin{bmatrix} 6 \\ 1 \\ 30 \end{bmatrix}$$

For the third eigenvalue, we have

$$Av_3 = v_3$$

Which gives three equations

$$2x + z = x$$
$$x - y = y$$
$$3x + 4z = z$$

Therefore we obtain from the first equation

$$z = -x$$

and from the second equation

$$x = 2y$$

So we can write the eigenvector as

$$v_3 = \begin{bmatrix} 2y \\ y \\ -2y \end{bmatrix}$$

Normalizing, we get

$$1 = v_3^\dagger v_3 = \begin{bmatrix} 2y^* & y^* & -2y^* \end{bmatrix} \begin{bmatrix} 2y \\ y \\ -2y \end{bmatrix} = 4\,|y|^2 + |y|^2 + 4\,|y|^2 = 9\,|y|^2$$

$$\Rightarrow |y|^2 = \tfrac{1}{9} \ or \ y = \tfrac{1}{3}$$

This allows us to write the third eigenvector as

$$v_3 = \begin{bmatrix} 2y \\ y \\ -2y \end{bmatrix} = \begin{bmatrix} \frac{2}{3} \\ \frac{1}{3} \\ -\frac{2}{3} \end{bmatrix} = \frac{1}{3} \begin{bmatrix} 2 \\ 1 \\ -2 \end{bmatrix}$$

The Eigenspace of an Operator A

The normalized eigenvectors of an operator A that belongs to a vector space V constitute a basis of V. If we are considering n-dimensional vectors in $\mathbb{R}^n$, then the normalized eigenvectors of A form a basis of R^n. Likewise, if we are working in $\mathbb{C}^n$, the normalized eigenvectors of an operator A form a basis of $\mathbb{C}^n$.

EXAMPLE 8-6
Consider the two-dimensional vector space $\mathbb{C}^2$. Find the normalized eigenvectors of

$$X = \begin{bmatrix} 0 & 1 \\ 1 & 0 \end{bmatrix}$$

and show that they constitute a basis.

SOLUTION 8-6
The characteristic equation is

$$0 = \det |X - \lambda I| = \det \left\| \begin{bmatrix} 0 & 1 \\ 1 & 0 \end{bmatrix} - \begin{bmatrix} \lambda & 0 \\ 0 & \lambda \end{bmatrix} \right\| = \det \begin{vmatrix} \lambda & 1 \\ 1 & \lambda \end{vmatrix}$$

$$\Rightarrow \lambda^2 - 1 = 0$$

This leads immediately to the eigenvalues

$$\lambda_1 = 1, \ \lambda_2 = -1$$

For the first eigenvalue we have

$$Xv_1 = v_1$$

Now

$$Xv_1 = \begin{bmatrix} 0 & 1 \\ 1 & 0 \end{bmatrix} \begin{bmatrix} x \\ y \end{bmatrix} = \begin{bmatrix} y \\ x \end{bmatrix}$$

Setting this equal to the eigenvector leads to $x = y$. So the eigenvector is

$$v_1 = \begin{bmatrix} x \\ x \end{bmatrix}$$

Normalizing we find

$$1 = v_1^\dagger v_1 = \begin{bmatrix} x^* & x^* \end{bmatrix} \begin{bmatrix} x \\ x \end{bmatrix} = x^*x + x^*x = |x|^2 + |x|^2 = 2\,|x|^2$$

$$\Rightarrow |x|^2 = \frac{1}{2} \text{ or } x = \frac{1}{\sqrt{2}}$$

and so the first eigenvector is

$$v_1 = \frac{1}{\sqrt{2}} \begin{bmatrix} 1 \\ 1 \end{bmatrix}$$

The second eigenvalue is $\lambda_2 = -1$. Setting $Xv_2 = -v_2$ gives

$$\begin{bmatrix} y \\ x \end{bmatrix} = \begin{bmatrix} -x \\ -y \end{bmatrix}$$

and so $y = -x$. Normalizing we find

$$1 = v_2^\dagger v_2 = \begin{bmatrix} x^* & -x^* \end{bmatrix} \begin{bmatrix} x \\ -x \end{bmatrix} = x^*x + (-x^*)(-x) = |x|^2 + |x|^2 = 2\,|x|^2$$

$$\Rightarrow |x|^2 = \frac{1}{2} \text{ or } x = \frac{1}{\sqrt{2}}$$

and so we have

$$v_2 = \begin{bmatrix} x \\ -x \end{bmatrix} = \frac{1}{\sqrt{2}} \begin{bmatrix} 1 \\ -1 \end{bmatrix}$$

We check to see if these eigenvectors satisfy the completeness relation:

$$v_1 v_1^\dagger + v_2 v_2^\dagger$$

$$= \begin{bmatrix} \frac{1}{\sqrt{2}} \\ \frac{1}{\sqrt{2}} \end{bmatrix} \begin{bmatrix} \frac{1}{\sqrt{2}} & \frac{1}{\sqrt{2}} \end{bmatrix} + \begin{bmatrix} \frac{1}{\sqrt{2}} \\ -\frac{1}{\sqrt{2}} \end{bmatrix} \begin{bmatrix} \frac{1}{\sqrt{2}} & -\frac{1}{\sqrt{2}} \end{bmatrix}$$

$$= \begin{bmatrix} \left(\frac{1}{\sqrt{2}}\right)\left(\frac{1}{\sqrt{2}}\right) & \left(\frac{1}{\sqrt{2}}\right)\left(\frac{1}{\sqrt{2}}\right) \\ \left(\frac{1}{\sqrt{2}}\right)\left(\frac{1}{\sqrt{2}}\right) & \left(\frac{1}{\sqrt{2}}\right)\left(\frac{1}{\sqrt{2}}\right) \end{bmatrix} + \begin{bmatrix} \left(\frac{1}{\sqrt{2}}\right)\left(\frac{1}{\sqrt{2}}\right) & \left(\frac{1}{\sqrt{2}}\right)\left(-\frac{1}{\sqrt{2}}\right) \\ \left(-\frac{1}{\sqrt{2}}\right)\left(\frac{1}{\sqrt{2}}\right) & \left(-\frac{1}{\sqrt{2}}\right)\left(-\frac{1}{\sqrt{2}}\right) \end{bmatrix}$$

$$= \begin{bmatrix} \frac{1}{2} & \frac{1}{2} \\ \frac{1}{2} & \frac{1}{2} \end{bmatrix} + \begin{bmatrix} \frac{1}{2} & -\frac{1}{2} \\ -\frac{1}{2} & \frac{1}{2} \end{bmatrix} = \begin{bmatrix} 1 & 0 \\ 0 & 1 \end{bmatrix} = I$$

Therefore the completeness relation is satisfied. Do the vectors span the space? We denote an arbitrary vector by

$$u = \begin{bmatrix} \alpha \\ \beta \end{bmatrix}$$

where α, β are complex numbers. To show that we can write this vector as a linear combination of the basis vectors of X, we expand it in terms of the basis vectors with complex numbers μ, ν:

$$u = \begin{bmatrix} \alpha \\ \beta \end{bmatrix} = \mu \frac{1}{\sqrt{2}} \begin{bmatrix} 1 \\ 1 \end{bmatrix} + \nu \frac{1}{\sqrt{2}} \begin{bmatrix} 1 \\ -1 \end{bmatrix}$$

This leads to the equations

$$\alpha = \frac{1}{\sqrt{2}} (\mu + \nu)$$

$$\beta = \frac{1}{\sqrt{2}} (\mu - \nu)$$

Adding these equations, we find that

$$\mu = \frac{(\alpha + \beta)}{\sqrt{2}}$$

Subtracting the second equation from the first gives

$$\nu = \frac{(\alpha - \beta)}{\sqrt{2}}$$

and so we can write any vector in $\mathbb{C}^2$ in terms of these basis vectors by writing

$$u = \begin{bmatrix} \alpha \\ \beta \end{bmatrix} = \frac{(\alpha + \beta)}{\sqrt{2}} \left(\frac{1}{\sqrt{2}} \begin{bmatrix} 1 \\ 1 \end{bmatrix} \right) + \frac{(\alpha - \beta)}{\sqrt{2}} \left(\frac{1}{\sqrt{2}} \begin{bmatrix} 1 \\ -1 \end{bmatrix} \right)$$

$$= \frac{(\alpha + \beta)}{\sqrt{2}} v_1 + \frac{(\alpha - \beta)}{\sqrt{2}} v_2$$

Therefore we conclude that the eigenvectors of X are complete and they span $\mathbb{C}^2$, therefore they constitute a basis of $\mathbb{C}^2$.

Similar Matrices

Two matrices A and B are *similar* if we can find a matrix S such that

$$B = S^{-1}AS$$

There is a theorem that states that if two matrices are similar, they have the same eigenvalues. This is helpful because we will see that if we can represent a matrix or operator in its own basis of eigenvectors, then that matrix will have a simple diagonal form with its eigenvalues along the diagonal.

EXAMPLE 8-7
Prove that similar matrices have the same eigenvalues.

SOLUTION 8-7
First recall that the determinant is just a number. So we can move determinants around in an expression at will. In addition, note that

$$\det |A^{-1}| = \frac{1}{\det |A|}$$

Now we form the characteristic equation

$$0 = \det |B - \lambda I| = \det \left| S^{-1}AS - \lambda I \right|$$

Now since $S^{-1}S = I$ and any matrix commutes with the identity matrix ($SI = IS$), we can write

$$\lambda I = \lambda \left(S^{-1}S \right) I = \lambda S^{-1}IS$$

Therefore we can rewrite the expression inside the determinant in the following way:

$$S^{-1}AS - \lambda I = S^{-1}AS - \lambda S^{-1}IS = \left(S^{-1}A - \lambda S^{-1}I\right)S = S^{-1}\left(A - \lambda I\right)S$$

and so the characteristic equation becomes

$$0 = \det \left|S^{-1}\left(A - \lambda I\right)S\right|$$

Now we invoke the product rule for determinants. This tells us that

$$\det|AB| = \det|A|\det|B|$$

This gives

$$0 = \det \left|S^{-1}\left(A - \lambda I\right)S\right| = \det \left|S^{-1}\right|\det|A - \lambda I|\det|S|$$

Remember, the determinant is just a number. So we can move these terms around and eliminate terms involving the similarity matrix S

$$\begin{aligned}
0 &= \det \left|S^{-1}\right|\det|A - \lambda I|\det|S| = \det \left|S^{-1}\right|\det|S|\det|A - \lambda I| \\
&= \det|A - \lambda I| \\
&\Rightarrow \det|B - \lambda I| = \det|A - \lambda I|
\end{aligned}$$

In other words, the similar matrices A and B have the same characteristic equation and therefore the same eigenvalues.

Diagonal Representations of an Operator

If a matrix is similar to a diagonal matrix, then we can write it in diagonal form. An important theorem tells us that a matrix that is a linear transformation T on a vector space V can be diagonalized if and only if the eigenvectors of T form a basis for V. Fortunately this is true for a large class of matrices, and it is true for *Hermitian* matrices that are important in physical applications.

You can check to see if the eigenvectors of a matrix form a basis by checking the following:

- Do the eigenvectors span the space; in other words, can you write any vector from the space in terms of the eigenvectors of the matrix?

- Are the eigenvectors linearly independent?
- Are they complete?

In the next chapter we will examine several special types of matrices. Note that

- The eigenvectors of a symmetric matrix form an orthonormal basis.
- The eigenvectors of a Hermitian matrix form an orthonormal basis.

Therefore symmetric and Hermitian matrices are diagonable.

If the matrix is diagonable, the eigenvectors of an operator or linear transformation allow us to write the matrix representation of that operator in a diagonal form. The diagonal representation of a matrix A is given by

$$A = \begin{bmatrix} \lambda_1 & 0 & \cdots & 0 \\ 0 & \lambda_2 & \cdots & \vdots \\ \vdots & \vdots & \ddots & 0 \\ 0 & 0 & \cdots & \lambda_n \end{bmatrix}$$

where λ_i are the eigenvalues of the matrix A. In this section we consider a special class of similar matrices that are related by *unitary transformations*. As we will see in the next chapter, a unitary matrix has the special property that

$$U^\dagger = U^{-1}$$

This makes it very easy to obtain the inverse of the matrix and to find the similarity relationship.

We obtain the diagonal form of a matrix by applying a unitary transformation. The unitary matrix U used in the transformation is constructed in the following way. The eigenvectors of the matrix A form the columns of the matrix U, i.e.,

$$U = \begin{bmatrix} v_1| & |\cdots| & v_n| \end{bmatrix}$$

The diagonal form of a matrix A, which we denote $\tilde{A}$, is found from

$$\tilde{A} = U^\dagger A U$$

This can be most easily seen with an example.

EXAMPLE 8-8

For the matrix X used in the previous example, use the eigenvectors to write down a unitary matrix U and show that it diagonalizes X, and that the diagonal entries are the eigenvalues of X.

SOLUTION 8-8

In the previous example we found that the eigenvectors of X were

$$v_1 = \frac{1}{\sqrt{2}} \begin{bmatrix} 1 \\ 1 \end{bmatrix} \quad \text{and} \quad v_2 = \frac{1}{\sqrt{2}} \begin{bmatrix} 1 \\ -1 \end{bmatrix}$$

The transformation matrix is constructed by setting the columns of the matrix equal to the eigenvectors:

$$U = \begin{bmatrix} v_1 & v_2 \end{bmatrix} = \begin{bmatrix} \frac{1}{\sqrt{2}} & \frac{1}{\sqrt{2}} \\ \frac{1}{\sqrt{2}} & -\frac{1}{\sqrt{2}} \end{bmatrix}$$

The Hermitian conjugate of this matrix is easy to compute; in fact we have

$$U^\dagger = \begin{bmatrix} \frac{1}{\sqrt{2}} & \frac{1}{\sqrt{2}} \\ \frac{1}{\sqrt{2}} & -\frac{1}{\sqrt{2}} \end{bmatrix}^\dagger = \begin{bmatrix} \frac{1}{\sqrt{2}} & \frac{1}{\sqrt{2}} \\ \frac{1}{\sqrt{2}} & -\frac{1}{\sqrt{2}} \end{bmatrix} = U$$

Now we apply this transformation to X:

$$U^\dagger X U = \begin{bmatrix} \frac{1}{\sqrt{2}} & \frac{1}{\sqrt{2}} \\ \frac{1}{\sqrt{2}} & -\frac{1}{\sqrt{2}} \end{bmatrix} \begin{bmatrix} 0 & 1 \\ 1 & 0 \end{bmatrix} \begin{bmatrix} \frac{1}{\sqrt{2}} & \frac{1}{\sqrt{2}} \\ \frac{1}{\sqrt{2}} & -\frac{1}{\sqrt{2}} \end{bmatrix}$$

$$= \begin{bmatrix} \frac{1}{\sqrt{2}} & \frac{1}{\sqrt{2}} \\ \frac{1}{\sqrt{2}} & -\frac{1}{\sqrt{2}} \end{bmatrix} \begin{bmatrix} \frac{1}{\sqrt{2}} & -\frac{1}{\sqrt{2}} \\ \frac{1}{\sqrt{2}} & \frac{1}{\sqrt{2}} \end{bmatrix}$$

$$= \begin{bmatrix} \frac{1}{2} + \frac{1}{2} & -\frac{1}{2} + \frac{1}{2} \\ \frac{1}{2} - \frac{1}{2} & -\frac{1}{2} - \frac{1}{2} \end{bmatrix}$$

$$= \begin{bmatrix} 1 & 0 \\ 0 & -1 \end{bmatrix}$$

In the previous example, we had found that the eigenvalues of X were ± 1. Therefore the diagonal matrix we found from this unitary transformation does have the eigenvalues of X along the diagonal.

The diagonal form of a matrix is a representation of that matrix with respect to its eigenbasis.

When two or more eignvectors share the same eigenvalue, we say that the eigenvalue is *degenerate*. The number of eigenvectors that have the same eigenvalue is the *degree* of degeneracy.

EXAMPLE 8-9
Diagonalize the matrix

$$A = \begin{bmatrix} 0 & 2 & 0 \\ 2 & 0 & 2 \\ 0 & 2 & 0 \end{bmatrix}$$

SOLUTION 8-9
Notice that this matrix is symmetric (we will see later it is also Hermitian) and so we know ahead of time that the eigenvectors constitute a basis. Solving the characteristic equation, we find

$$0 = \det \left| \begin{bmatrix} 0 & 2 & 0 \\ 2 & 0 & 2 \\ 0 & 2 & 0 \end{bmatrix} - \begin{bmatrix} \lambda & 0 & 0 \\ 0 & \lambda & 0 \\ 0 & 0 & \lambda \end{bmatrix} \right| = \det \begin{vmatrix} -\lambda & 2 & 0 \\ 2 & -\lambda & 2 \\ 0 & 2 & -\lambda \end{vmatrix}$$

This leads to the eigenvalues $\{0, -2\sqrt{2}, 2\sqrt{2}\}$ (exercise). For the first eigenvalue we have

$$Av_1 = 0$$

This leads to the equations

$$2y = 0, \Rightarrow y = 0$$
$$2x + 2z = 0, \Rightarrow z = -x$$

Therefore the eigenvector can be written as

$$v_1 = \begin{bmatrix} x \\ 0 \\ -x \end{bmatrix}$$

Normalization gives

$$1 = \begin{bmatrix} x^* & 0 & -x^* \end{bmatrix} \begin{bmatrix} x \\ 0 \\ -x \end{bmatrix} = 2\,|x|^2$$

Therefore we can take $x = \frac{1}{\sqrt{2}}$, and the first eigenvector is

$$v_1 = \frac{1}{\sqrt{2}} \begin{bmatrix} 1 \\ 0 \\ -1 \end{bmatrix}$$

For the second eigenvalue we have

$$Av_2 = -2\sqrt{2}v_2$$

This leads to the equations

$$2y = -2\sqrt{2}x$$
$$2x + 2z = -2\sqrt{2}y$$
$$2y = -2\sqrt{2}z$$

A little manipulation shows that

$$y = -\sqrt{2}x, \; z = -\frac{1}{\sqrt{2}}y = x$$

and so we have

$$v_2 = \begin{bmatrix} x \\ -\sqrt{2}x \\ x \end{bmatrix}$$

Normalization gives

$$1 = v_2^{\dagger}v_2 = \begin{bmatrix} x^* & -\sqrt{2}x^* & x^* \end{bmatrix} \begin{bmatrix} x \\ -\sqrt{2}x \\ x \end{bmatrix} = |x|^2 + 2\,|x|^2 + |x|^2 = 4\,|x|^2$$

$$\Rightarrow |x|^2 = \frac{1}{4} \text{ or } x = \frac{1}{2}$$

and so the normalized eigenvector is

$$v_2 = \begin{bmatrix} x \\ -\sqrt{2}x \\ x \end{bmatrix} = \frac{1}{2}\begin{bmatrix} 1 \\ -\sqrt{2} \\ 1 \end{bmatrix}$$

The third eigenvalue equation is

$$Av_3 = 2\sqrt{2}v_3$$

A similar procedure shows that the third eigenvector is

$$v_3 = \frac{1}{2}\begin{bmatrix} 1 \\ \sqrt{2} \\ 1 \end{bmatrix}$$

The unitary matrix that diagonalizes A is found by setting its columns equal to the normalized eigenvectors of A

$$U = \begin{bmatrix} v_1 & v_2 & v_3 \end{bmatrix} = \begin{bmatrix} \frac{1}{\sqrt{2}} & \frac{1}{2} & \frac{1}{2} \\ 0 & -\frac{\sqrt{2}}{2} & \frac{\sqrt{2}}{2} \\ -\frac{1}{\sqrt{2}} & \frac{1}{2} & \frac{1}{2} \end{bmatrix}$$

The inverse of this matrix is found from $U^\dagger$, which is

$$U^\dagger = \begin{bmatrix} \frac{1}{\sqrt{2}} & 0 & -\frac{1}{\sqrt{2}} \\ \frac{1}{2} & -\frac{\sqrt{2}}{2} & \frac{1}{2} \\ \frac{1}{2} & \frac{\sqrt{2}}{2} & \frac{1}{2} \end{bmatrix}$$

Now we apply the transformation to the matrix A:

$$U^\dagger A U = \begin{bmatrix} \frac{1}{\sqrt{2}} & 0 & -\frac{1}{\sqrt{2}} \\ \frac{1}{2} & -\frac{\sqrt{2}}{2} & \frac{1}{2} \\ \frac{1}{2} & \frac{\sqrt{2}}{2} & \frac{1}{2} \end{bmatrix} \begin{bmatrix} 0 & 2 & 0 \\ 2 & 0 & 2 \\ 0 & 2 & 0 \end{bmatrix} \begin{bmatrix} \frac{1}{\sqrt{2}} & \frac{1}{2} & \frac{1}{2} \\ 0 & -\frac{\sqrt{2}}{2} & \frac{\sqrt{2}}{2} \\ -\frac{1}{\sqrt{2}} & \frac{1}{2} & \frac{1}{2} \end{bmatrix}$$

$$= \begin{bmatrix} \frac{1}{\sqrt{2}} & 0 & -\frac{1}{\sqrt{2}} \\ \frac{1}{2} & -\frac{\sqrt{2}}{2} & \frac{1}{2} \\ \frac{1}{2} & \frac{\sqrt{2}}{2} & \frac{1}{2} \end{bmatrix} \begin{bmatrix} 0 & -\sqrt{2} & \sqrt{2} \\ 0 & 2 & 2 \\ 0 & -\sqrt{2} & \sqrt{2} \end{bmatrix}$$

$$= \begin{bmatrix} 0 & 0 & 0 \\ 0 & -2\sqrt{2} & 0 \\ 0 & 0 & 2\sqrt{2} \end{bmatrix}$$

The Trace and Determinant and Eigenvalues

The trace of a matrix is equal to the sum of its eigenvalues:

$$\mathrm{tr}\,(A) = \sum \lambda_i$$

whereas the determinant of a matrix is equal to the product of its eigenvalues:

$$\det |A| = \prod \lambda_i$$

EXAMPLE 8-10
Find the trace and determinant of

$$A = \begin{bmatrix} 5 & 2 \\ 9 & 2 \end{bmatrix}$$

Using its eigenvalues.

SOLUTION 8-10
Earlier we found that the eigenvalues of the matrix are

$$\lambda_1 = 8, \quad \lambda_2 = -1$$

The trace of the matrix can be found from the sum of the diagonal elements:

$$\mathrm{tr}\,(A) = \mathrm{tr}\left(\begin{bmatrix} 5 & 2 \\ 9 & 2 \end{bmatrix} \right) = 5 + 2 = 7$$

or, from the sum of the eigenvalues:

$$\mathrm{tr}\,(A) = \lambda_1 + \lambda_2 = 8 - 1 = 7$$

The determinant is

$$\det |A| = \det \left| \begin{bmatrix} 5 & 2 \\ 9 & 2 \end{bmatrix} \right| = 10 - 18 = -8$$

or, from the product of the eigenvalues,

$$\det |A| = (\lambda_1)(\lambda_2) = (8)(-1) = -8$$

Quiz

1. Find the characteristic polynomial and show that the eigenvalues for the matrix

$$A = \begin{bmatrix} -1 & 4 \\ 1 & 2 \end{bmatrix}$$

are $\{-2, 3\}$.

2. Find the eigenvalues of the matrix

$$B = \begin{bmatrix} 4 & 0 & -1 \\ 0 & 2 & 8 \\ -1 & 0 & 1 \end{bmatrix}$$

3. Show that the eigenvalues of the matrix

$$Z = \begin{bmatrix} 1 & 0 \\ 0 & -1 \end{bmatrix}$$

are ± 1.

4. Find the eigenvectors of

$$Z = \begin{bmatrix} 1 & 0 \\ 0 & -1 \end{bmatrix}$$

5. Are the eigenvectors of Z found in the previous problem a basis for $\mathbb{C}^2$?

6. Show that the matrix

$$A = \begin{bmatrix} 3 & 1 & 0 \\ 0 & 3 & 1 \\ 3 & -7 & 8 \end{bmatrix}$$

has degenerate eigenvalues $\{4, 4, 6\}$. What is the degree of degeneracy? Find the eigenvectors of the matrix.

7. Show that the eigenvalues of

$$A = \begin{bmatrix} 0 & 2 & 0 \\ 2 & 0 & 2 \\ 0 & 2 & 0 \end{bmatrix}$$

are $\left\{0, -2\sqrt{2}, 2\sqrt{2}\right\}$.

8. Verify that the matrix

$$U = \begin{bmatrix} \frac{1}{\sqrt{2}} & \frac{1}{2} & \frac{1}{2} \\ 0 & -\frac{\sqrt{2}}{2} & \frac{\sqrt{2}}{2} \\ -\frac{1}{\sqrt{2}} & \frac{1}{2} & \frac{1}{2} \end{bmatrix}$$

is unitary.

9. Are the eigenvectors of

$$A = \begin{bmatrix} 0 & 2 & 0 \\ 2 & 0 & 2 \\ 0 & 2 & 0 \end{bmatrix}$$

a basis of $\mathbb{R}^3$?

10. Verify the Cayley-Hamilton theorem for the matrix

$$X = \begin{bmatrix} 0 & 1 \\ 1 & 0 \end{bmatrix}$$

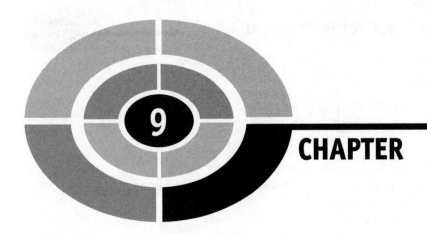

Special Matrices

In this chapter we give an overview of matrices that have special properties. We begin by considering symmetric matrices.

Symmetric and Skew-Symmetric Matrices

An $n \times n$ matrix is *symmetric* if it is equal to its transpose, i.e.,

$$A^T = A$$

The sum of two symmetric matrices is also symmetric. Let A and B be symmetric matrices so that

$$A^T = A, \quad B^T = B$$

Then we have

$$(A + B)^T = A^T + B^T = A + B$$

The product of two symmetric matrices may or may not be symmetric. Again letting A and B be symmetric matrices, the transpose of the product is

$$(AB)^T = B^T A^T = BA$$

For the product of two symmetric matrices to be symmetric, we must have

$$(AB)^T = AB$$

Therefore we see that the product of two symmetric matrices is symmetric only if the matrices A and B commute, meaning that

$$AB = BA$$

We can write any symmetric matrix S as the sum of some other matrix A and its transpose. That is

$$S = \frac{1}{2}\left(A + A^T\right)$$

Then

$$S^T = \frac{1}{2}\left(A + A^T\right)^T = \frac{1}{2}\left[A^T + \left(A^T\right)^T\right] = \frac{1}{2}\left(A^T + A\right) = \frac{1}{2}\left(A + A^T\right) = S$$

Another way of looking at this is that we can write any matrix A as a symmetric matrix by forming this sum.

EXAMPLE 9-1
Let

$$A = \begin{bmatrix} 2 & -4 \\ 3 & -1 \end{bmatrix}$$

Use it to construct a symmetric matrix.

SOLUTION 9-1
We compute the transpose

$$A^T = \begin{bmatrix} 2 & -4 \\ 3 & -1 \end{bmatrix}^T = \begin{bmatrix} 2 & 3 \\ -4 & -1 \end{bmatrix}$$

Clearly $A \neq A^T$ and so this particular matrix is not symmetric. Now we use it to construct a symmetric matrix

$$A + A^T = \begin{bmatrix} 2 & -4 \\ 3 & -1 \end{bmatrix} + \begin{bmatrix} 2 & 3 \\ -4 & -1 \end{bmatrix} = \begin{bmatrix} 4 & -1 \\ -1 & -2 \end{bmatrix}$$

$$S = \frac{1}{2}(A + A^T) = \frac{1}{2}\begin{bmatrix} 4 & -1 \\ -1 & -2 \end{bmatrix} = \begin{bmatrix} 2 & -\frac{1}{2} \\ -\frac{1}{2} & -1 \end{bmatrix}$$

$$S^T = \begin{bmatrix} 2 & -\frac{1}{2} \\ -\frac{1}{2} & -1 \end{bmatrix}^T = \begin{bmatrix} 2 & -\frac{1}{2} \\ -\frac{1}{2} & -1 \end{bmatrix} = S$$

Therefore we have constructed a symmetric matrix.

EXAMPLE 9-2
Suppose that

$$A = \begin{bmatrix} 2 & 1 \\ 1 & 4 \end{bmatrix} \quad \text{and} \quad B = \begin{bmatrix} -8 & 3 \\ 3 & 1 \end{bmatrix}$$

Are these matrices symmetric? Is their product symmetric?

SOLUTION 9-2
We immediately see the matrices are symmetric

$$A = \begin{bmatrix} 2 & 1 \\ 1 & 4 \end{bmatrix} \Rightarrow A^T = \begin{bmatrix} 2 & 1 \\ 1 & 4 \end{bmatrix} = A$$

$$B = \begin{bmatrix} -8 & 3 \\ 3 & 1 \end{bmatrix} \Rightarrow B^T = \begin{bmatrix} -8 & 3 \\ 3 & 1 \end{bmatrix} = B$$

We calculate the product

$$AB = \begin{bmatrix} 2 & 1 \\ 1 & 4 \end{bmatrix}\begin{bmatrix} -8 & 3 \\ 3 & 1 \end{bmatrix} = \begin{bmatrix} (2)(-8)+(1)(3) & (2)(3)+(1)(1) \\ (1)(-8)+(4)(3) & (1)(3)+(4)(1) \end{bmatrix}$$

$$= \begin{bmatrix} -13 & 7 \\ 4 & 7 \end{bmatrix}$$

Since the off-diagonal entries are not equal, we see this matrix is not symmetric. We calculate the transpose to verify this explicitly:

$$(AB)^T = \begin{bmatrix} -13 & 7 \\ 4 & 7 \end{bmatrix}^T = \begin{bmatrix} -13 & 4 \\ 7 & 7 \end{bmatrix} \neq AB$$

Another way to see this is to calculate

$$BA = \begin{bmatrix} -8 & 3 \\ 3 & 1 \end{bmatrix} \begin{bmatrix} 2 & 1 \\ 1 & 4 \end{bmatrix} = \begin{bmatrix} (-8)(2)+(3)(1) & (-8)(1)+(3)(4) \\ (3)(2)+(1)(1) & (3)(1)+(1)(4) \end{bmatrix}$$
$$= \begin{bmatrix} -13 & 4 \\ 7 & 7 \end{bmatrix}$$

We see that the matrices do not commute, thus the product cannot be symmetric.

SKEW SYMMETRY

A *skew-symmetric* matrix K has the property that

$$K = -K^T$$

It is possible to use any matrix A to create a skew-symmetric matrix by writing

$$K = \frac{1}{2}\left(A - A^T\right)$$
$$\Rightarrow K^T = \frac{1}{2}\left(A - A^T\right)^T = \frac{1}{2}\left(A^T - A\right) = -\frac{1}{2}\left(A - A^T\right) = -K$$

EXAMPLE 9-3
What can you say, if anything, about the diagonal elements of a skew-symmetric matrix? To do the proof, consider the 3×3 case.

SOLUTION 9-3
We write an arbitrary matrix

$$A = \begin{pmatrix} a_{11} & a_{12} & a_{13} \\ a_{21} & a_{22} & a_{23} \\ a_{31} & a_{32} & a_{33} \end{pmatrix}$$

The transpose is

$$A^T = \begin{pmatrix} a_{11} & a_{21} & a_{31} \\ a_{12} & a_{22} & a_{32} \\ a_{13} & a_{23} & a_{33} \end{pmatrix}$$

To be skew-symmetric, we must have $A = -A^T$. This leads to the following relationships:

$$a_{12} = -a_{21}, \quad a_{13} = -a_{31}, \quad a_{23} = -a_{32}$$

It must also be true that

$$a_{11} = -a_{11}, \quad a_{22} = -a_{22}, \quad a_{33} = -a_{33}$$

This condition can be met only if

$$a_{11} = a_{22} = a_{33} = 0$$

Therefore we conclude that the diagonal elements of a skew-symmetric matrix must be zero.

EXAMPLE 9-4
Let A and B be skew-symmetric matrices. Are the sum and product of these matrices skew-symmetric?

SOLUTION 9-4
We have

$$A^T = -A$$
$$B^T = -B$$

Therefore

$$(A + B)^T = A^T + B^T = -A - B = -(A + B)$$

So we conclude that the sum of two skew-symmetric matrices is skew-symmetric. For the product we have

$$(AB)^T = B^T A^T = (-B)(-A) = BA$$

The product would be skew symmetric if

$$(AB)^T = -AB$$

Therefore we must have

$$AB = -BA$$

or

$$AB + BA = 0$$

That is, the matrices must *anticommute*. The sum $AB + BA$ is called the anti-commutator and is written as

$$\{A, B\} = AB + BA$$

Hermitian Matrices

We now consider a special type of matrix with complex elements called a *Hermitian* matrix. A Hermitian matrix has the property that

$$A = A^\dagger$$

Some properties of the Hermitian conjugate operation to note are that

$$\left(A^\dagger\right)^\dagger = A$$

and

$$(A + B)^\dagger = A^\dagger + B^\dagger$$
$$(AB)^\dagger = B^\dagger A^\dagger$$

EXAMPLE 9-5
Are the matrices

$$Y = \begin{bmatrix} 0 & -i \\ i & 0 \end{bmatrix} \quad \text{and} \quad C = \begin{bmatrix} 3i & 0 & 2i \\ 0 & 4 & 6 \\ -2i & 1 & 0 \end{bmatrix}$$

Hermitian?

SOLUTION 9-5

We compute the Hermitian conjugate of each matrix, beginning by calculating the transpose of Y:

$$Y^T = \begin{bmatrix} 0 & -i \\ i & 0 \end{bmatrix}^T = \begin{bmatrix} 0 & i \\ -i & 0 \end{bmatrix}$$

Now we take the complex conjugate of each component, by letting $i \to -i$:

$$Y^\dagger = \begin{bmatrix} 0 & i \\ -i & 0 \end{bmatrix}^* = \begin{bmatrix} 0 & -i \\ i & 0 \end{bmatrix} = Y$$

Therefore Y is a Hermitian matrix. The transpose of C is

$$C^T = \begin{bmatrix} 3i & 0 & 2i \\ 0 & 4 & 6 \\ -2i & 1 & 0 \end{bmatrix}^T = \begin{bmatrix} 3i & 0 & -2i \\ 0 & 4 & 1 \\ 2i & 6 & 0 \end{bmatrix}$$

Taking the complex conjugate, we find the Hermitian conjugate to be

$$C^\dagger = \begin{bmatrix} 3i & 0 & -2i \\ 0 & 4 & 1 \\ 2i & 6 & 0 \end{bmatrix}^* = \begin{bmatrix} -3i & 0 & 2i \\ 0 & 4 & 1 \\ -2i & 6 & 0 \end{bmatrix} \neq C$$

and so the matrix C is not Hermitian.

Some important facts about Hermitian matrices are:

- The diagonal elements of a Hermitian matrix are real numbers
- Hermitian matrices have real eigenvalues
- The eigenvectors of a Hermitian matrix are orthogonal. In fact they constitute a basis.

EXAMPLE 9-6

Prove that a 3×3 Hermitian matrix must have real elements along the diagonal. What can be said about the off-diagonal elements?

SOLUTION 9-6
We set

$$A = \begin{pmatrix} a_{11} & a_{12} & a_{13} \\ a_{21} & a_{22} & a_{23} \\ a_{31} & a_{32} & a_{33} \end{pmatrix}$$

Therefore we have

$$A^\dagger = \begin{pmatrix} a_{11}^* & a_{21}^* & a_{31}^* \\ a_{12}^* & a_{22}^* & a_{32}^* \\ a_{13}^* & a_{23}^* & a_{33}^* \end{pmatrix}$$

We consider the off-diagonal elements first. For this matrix to be Hermitian, it must be the case that

$$a_{21}^* = a_{12}, \quad a_{31}^* = a_{13}, \quad a_{32}^* = a_{23}$$

This is also true for the complex conjugates of these relations. In addition, we need to have

$$a_{11}^* = a_{11}, \quad a_{22}^* = a_{22}, \quad a_{33}^* = a_{33}$$

This can be true only if

$$a_{11}, \quad a_{22}, \quad a_{33}$$

are real numbers.

EXAMPLE 9-7
Is the matrix

$$B = \begin{bmatrix} 4 & 0 & 0 \\ 0 & 2 & i \\ 0 & -i & 1 \end{bmatrix}$$

Hermitian? Does it have real eigenvalues?

SOLUTION 9-7
The transpose is

$$B^T = \begin{bmatrix} 4 & 0 & 0 \\ 0 & 2 & i \\ 0 & -i & 1 \end{bmatrix}^T = \begin{bmatrix} 4 & 0 & 0 \\ 0 & 2 & -i \\ 0 & i & 1 \end{bmatrix}$$

Taking the complex conjugate

$$B^\dagger = \begin{bmatrix} 4 & 0 & 0 \\ 0 & 2 & -i \\ 0 & i & 1 \end{bmatrix}^* = \begin{bmatrix} 4 & 0 & 0 \\ 0 & 2 & i \\ 0 & -i & 1 \end{bmatrix} = B$$

Therefore the matrix is Hermitian. The characteristic equation is

$$\det|B - \lambda I| = \det \left| \begin{bmatrix} 4 & 0 & 0 \\ 0 & 2 & -i \\ 0 & i & 1 \end{bmatrix} - \begin{bmatrix} \lambda & 0 & 0 \\ 0 & \lambda & 0 \\ 0 & 0 & \lambda \end{bmatrix} \right| = \det \begin{vmatrix} 4 - \lambda & 0 & 0 \\ 0 & 2 - \lambda & -i \\ 0 & i & 1 - \lambda \end{vmatrix}$$

$$= (4 - \lambda) \det \begin{vmatrix} 2 - \lambda & -i \\ i & 1 - \lambda \end{vmatrix}$$

$$= (4 - \lambda)\left[(2 - \lambda)(1 - \lambda) - 1\right]$$

$$= (4 - \lambda)\left[\lambda^2 - 3\lambda + 1\right] = 0$$

We see from the first term in the product that the first eigenvalue is

$$(4 - \lambda) = 0, \quad \lambda_1 = 4$$

Which is a real number. The other term gives

$$\lambda^2 - 3\lambda + 1 = 0$$

$$\Rightarrow \lambda_{2,3} = \frac{3 \pm \sqrt{9 - 4}}{2} = \frac{3 \pm \sqrt{5}}{2}$$

These are both real numbers. Therefore B, which is a Hermitian matrix, has real eigenvalues as expected.

ANTI-HERMITIAN MATRICES

An *anti-Hermitian* matrix A is one that satisfies

$$A^\dagger = -A$$

Anti-Hermitian matrices have purely imaginary elements along the diagonal and have imaginary eigenvalues.

EXAMPLE 9-8
Construct Hermitian and anti-Hermitian matrices out of an arbitrary matrix A.

SOLUTION 9-8

To construct a Hermitian matrix, we consider the sum

$$B = A + A^\dagger$$

Since $\left(A^\dagger\right)^\dagger = A$, we find that

$$B^\dagger = \left(A + A^\dagger\right)^\dagger = A^\dagger + \left(A^\dagger\right)^\dagger = A^\dagger + A = A + A^\dagger = B$$

Therefore B is a Hermitian matrix. Now consider the sum

$$C = i\left(A + A^\dagger\right)$$

The Hermitian conjugate of this matrix is

$$C^\dagger = \left[i\left(A + A^\dagger\right)\right]^\dagger = -i\left(A^\dagger + A\right) = -C$$

This is true because the Hermitian conjugate of a *number* is given by the complex conjugate; therefore, $i \to -i$.

Orthogonal Matrices

An *orthogonal* matrix is an $n \times n$ matrix whose columns or rows form an orthonormal basis for $\mathbb{R}^n$. We have already seen the simplest orthogonal matrix, the identity matrix. Consider the identity matrix in 3 dimensions:

$$I = \begin{bmatrix} 1 & 0 & 0 \\ 0 & 1 & 0 \\ 0 & 0 & 1 \end{bmatrix}$$

The columns are

$$v_1 = \begin{bmatrix} 1 \\ 0 \\ 0 \end{bmatrix}, \quad v_2 = \begin{bmatrix} 0 \\ 1 \\ 0 \end{bmatrix}, \quad v_3 = \begin{bmatrix} 0 \\ 0 \\ 1 \end{bmatrix}$$

It is immediately obvious that these columns are orthonormal. Consider the first column

$$(v_1, v_1) = \begin{bmatrix} 1 & 0 & 0 \end{bmatrix} \begin{bmatrix} 1 \\ 0 \\ 0 \end{bmatrix} = (1)(1) + (0)(0) + (0)(0) = 1$$

and we have

$$(v_1, v_2) = \begin{bmatrix} 1 & 0 & 0 \end{bmatrix} \begin{bmatrix} 0 \\ 1 \\ 0 \end{bmatrix} = (1)(0) + (0)(1) + (0)(0) = 0$$

and so on. An important property of an orthogonal matrix P is that

$$P^T = P^{-1}$$

EXAMPLE 9-9
Is the matrix

$$B = \begin{bmatrix} 1 & -1 \\ 1 & 1 \end{bmatrix}$$

orthogonal?

SOLUTION 9-9
We have

$$B^T = \begin{bmatrix} 1 & -1 \\ 1 & 1 \end{bmatrix}^T = \begin{bmatrix} 1 & 1 \\ -1 & 1 \end{bmatrix}$$

Therefore

$$BB^T = \begin{bmatrix} 1 & -1 \\ 1 & 1 \end{bmatrix} \begin{bmatrix} 1 & 1 \\ -1 & 1 \end{bmatrix} = \begin{bmatrix} (1)(1) + (-1)(-1) & (1)(1) + (-1)(1) \\ (1)(1) + (1)(-1) & (1)(1) + (1)(1) \end{bmatrix}$$

$$= \begin{bmatrix} 1+1 & 1-1 \\ 1-1 & 1+1 \end{bmatrix}$$

$$= \begin{bmatrix} 2 & 0 \\ 0 & 2 \end{bmatrix} = 2\begin{bmatrix} 1 & 0 \\ 0 & 1 \end{bmatrix} = 2I$$

The answer is that this matrix is not quite orthogonal. Looking at the inner product of the columns, we have

$$(v_1, v_2) = \begin{bmatrix} 1 & 1 \end{bmatrix} \begin{bmatrix} 1 \\ -1 \end{bmatrix} = (1)(1) + (1)(-1) = 1 - 1 = 0$$

So the columns are orthogonal (you can verify the rows are as well). However

$$(v_1, v_1) = \begin{bmatrix} 1 & 1 \end{bmatrix} \begin{bmatrix} 1 \\ 1 \end{bmatrix} = (1)(1) + (1)(1) = 1 + 1 = 2$$

So we see that the columns are not normalized. It is a simple matter to see that the matrix

$$\tilde{B} = \frac{1}{\sqrt{2}} B = \begin{bmatrix} \frac{1}{\sqrt{2}} & -\frac{1}{\sqrt{2}} \\ \frac{1}{\sqrt{2}} & \frac{1}{\sqrt{2}} \end{bmatrix}$$

is orthogonal. The product of this matrix with the transpose does give the identity and the columns (rows) are normalized.

EXAMPLE 9-10
Are the following transformations orthogonal?

$$T(x, y, z) = (x - z, x + y, z)$$
$$L(x, y, z) = (z, x, y)$$

SOLUTION 9-10
We represent the first transformation as the matrix

$$T = \begin{bmatrix} 1 & 0 & -1 \\ 1 & 1 & 0 \\ 0 & 0 & 1 \end{bmatrix}$$

We check by acting this operator on a column vector

$$Tv = \begin{bmatrix} 1 & 0 & -1 \\ 1 & 1 & 0 \\ 0 & 0 & 1 \end{bmatrix} \begin{bmatrix} x \\ y \\ z \end{bmatrix} = \begin{bmatrix} (1)(x) + (0)(y) - (1)(z) \\ (1)(x) + (1)(y) + (0)(z) \\ (0)(x) + (0)(y) + (1)(z) \end{bmatrix} = \begin{bmatrix} x - z \\ x + y \\ z \end{bmatrix}$$

The transpose of this matrix is

$$T^T = \begin{bmatrix} 1 & 0 & -1 \\ 1 & 1 & 0 \\ 0 & 0 & 1 \end{bmatrix}^T = \begin{bmatrix} 1 & 1 & 0 \\ 0 & 1 & 0 \\ -1 & 0 & 1 \end{bmatrix}$$

We find that

$$TT^T = \begin{bmatrix} 1 & 0 & -1 \\ 1 & 1 & 0 \\ 0 & 0 & 1 \end{bmatrix} \begin{bmatrix} 1 & 1 & 0 \\ 0 & 1 & 0 \\ -1 & 0 & 1 \end{bmatrix} = \begin{bmatrix} 2 & 1 & -1 \\ 1 & 2 & 0 \\ -1 & 0 & 1 \end{bmatrix} \neq I$$

Therefore the first transformation is not orthogonal. For the second transformation, we can write this as the matrix

$$L = \begin{bmatrix} 0 & 0 & 1 \\ 1 & 0 & 0 \\ 0 & 1 & 0 \end{bmatrix}$$

(check). This matrix can be obtained from the identity matrix by a series of elementary row operations. It is a fact that such a matrix is orthogonal. We check it explicitly. The transpose is

$$L^T = \begin{bmatrix} 0 & 0 & 1 \\ 1 & 0 & 0 \\ 0 & 1 & 0 \end{bmatrix}^T = \begin{bmatrix} 0 & 1 & 0 \\ 0 & 0 & 1 \\ 1 & 0 & 0 \end{bmatrix}$$

and we have

$$LL^T = \begin{bmatrix} 0 & 0 & 1 \\ 1 & 0 & 0 \\ 0 & 1 & 0 \end{bmatrix} \begin{bmatrix} 0 & 1 & 0 \\ 0 & 0 & 1 \\ 1 & 0 & 0 \end{bmatrix} = \begin{bmatrix} 1 & 0 & 0 \\ 0 & 1 & 0 \\ 0 & 0 & 1 \end{bmatrix} = I$$

Therefore the transformation L is orthogonal. You can check that the rows and columns of the matrix are orthonormal.

ORTHOGONAL MATRICES AND ROTATIONS

A 2×2 orthogonal matrix can be written in the form

$$A = \begin{bmatrix} \cos\phi & \sin\phi \\ \sin\phi & -\cos\phi \end{bmatrix}$$

for some angle ϕ. Now the transpose of this matrix is

$$A^T = \begin{bmatrix} \cos\phi & \sin\phi \\ \sin\phi & -\cos\phi \end{bmatrix}^T = \begin{bmatrix} \cos\phi & \sin\phi \\ \sin\phi & -\cos\phi \end{bmatrix} = A$$

But notice that

$$AA^T = \begin{bmatrix} \cos\phi & \sin\phi \\ \sin\phi & -\cos\phi \end{bmatrix} \begin{bmatrix} \cos\phi & \sin\phi \\ \sin\phi & -\cos\phi \end{bmatrix}$$

$$= \begin{bmatrix} \cos^2\phi & +\sin^2\phi & \cos\phi\sin\phi & -\sin\phi\cos\phi \\ \sin\phi\cos\phi & -\cos\phi\sin\iota & \sin^2\phi & +\cos^2\phi \end{bmatrix}$$

$$= \begin{bmatrix} 1 & 0 \\ 0 & 1 \end{bmatrix}$$

Consider the inner product between the columns

$$(v_1, v_2) = \begin{bmatrix} \cos\phi & \sin\phi \end{bmatrix} \begin{bmatrix} \sin\phi \\ -\cos\phi \end{bmatrix} = \cos\phi\sin\phi - \sin\phi\cos\phi = 0$$

$$(v_1, v_1) = \begin{bmatrix} \cos\phi & \sin\phi \end{bmatrix} \begin{bmatrix} \cos\phi \\ \sin\phi \end{bmatrix} = \cos^2\phi + \sin^2\phi = 1$$

$$(v_2, v_2) = \begin{bmatrix} \sin\phi & -\cos\phi \end{bmatrix} \begin{bmatrix} \sin\phi \\ -\cos\phi \end{bmatrix} = \sin^2\phi + \cos^2\phi = 1$$

It is easy to verify that the inner products among the rows works out the same way. We have a matrix that when multiplied by the transpose gives the identity, and which has orthonormal rows and columns. Therefore the rotation matrix is orthogonal.

This matrix is called the rotation matrix because of its action on a vector in the plane. We have

$$AX = \begin{bmatrix} \cos\phi & \sin\phi \\ \sin\phi & -\cos\phi \end{bmatrix} \begin{bmatrix} x \\ y \end{bmatrix} = \begin{bmatrix} x\cos\phi + y\sin\phi \\ x\sin\phi - y\cos\phi \end{bmatrix}$$

A rotation in the plane can be visualized as shown in Fig. 9-1.

An examination of the figure shows that the rotation transforms the coordinates in the same way as the matrix A does.

Rotations in 3 dimensions can be taken about the x, y, and z axes, respectively. These rotations are represented by the matrices

$$R_x = \begin{bmatrix} 1 & 0 & 0 \\ 0 & \cos\phi & -\sin\phi \\ 0 & \sin\phi & \cos\phi \end{bmatrix}$$

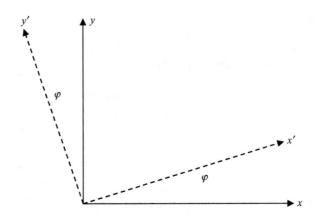

Fig. 9-1. A rotation in the plane.

$$R_y = \begin{bmatrix} \cos\phi & 0 & \sin\phi \\ 0 & 1 & 0 \\ -\sin\phi & 0 & \cos\phi \end{bmatrix}$$

$$R_z = \begin{bmatrix} \cos\phi & -\sin\phi & 0 \\ \sin\phi & \cos\phi & 0 \\ 0 & 0 & 1 \end{bmatrix}$$

Unitary Matrices

We have already come across unitary matrices in our studies. A *unitary matrix* is a complex generalization of an orthogonal matrix. Unitary matrices are characterized by the following property:

$$UU^\dagger = U^\dagger U = I$$

In other words, the Hermitian conjugate of a unitary matrix is its inverse:

$$U^\dagger = U^{-1}$$

Unitary matrices play a central role in the study of quantum theory. The "Pauli Matrices"

$$X = \begin{bmatrix} 0 & 1 \\ 1 & 0 \end{bmatrix}, \quad Y = \begin{bmatrix} 0 & -i \\ i & 0 \end{bmatrix}, \quad Z = \begin{bmatrix} 1 & 0 \\ 0 & -1 \end{bmatrix}$$

are all both Hermitian and unitary.

EXAMPLE 9-11

Verify that

$$Y = \begin{bmatrix} 0 & -i \\ i & 0 \end{bmatrix}$$

is unitary.

SOLUTION 9-11

The transpose is

$$Y^T = \begin{bmatrix} 0 & i \\ -i & 0 \end{bmatrix}$$

Therefore the Hermitian conjugate is

$$Y^\dagger = \begin{bmatrix} 0 & -i \\ i & 0 \end{bmatrix} = Y$$

We have found that Y is Hermitian. Now we check to see if it is unitary

$$YY^\dagger = \begin{bmatrix} 0 & -i \\ i & 0 \end{bmatrix}\begin{bmatrix} 0 & -i \\ i & 0 \end{bmatrix} = \begin{bmatrix} (-i)(i) & 0 \\ 0 & (i)(-i) \end{bmatrix} = \begin{bmatrix} 1 & 0 \\ 0 & 1 \end{bmatrix}$$

and in fact the matrix is unitary.

Unitary matrices have eigenvalues that are complex numbers with modulus 1. We have already seen that the eigenvectors of a Hermitian matrix can be used to construct a unitary matrix that transforms the Hermitian matrix into a diagonal one. It is also true that a unitary matrix can also be constructed to perform a change of basis. For simplicity we consider a three-dimensional space. If we represent one basis by u_i and a second basis by v_i then the change of basis matrix is

$$\begin{bmatrix} (v_1, u_1) & (v_1, u_2) & (v_1, u_3) \\ (v_2, u_1) & (v_2, u_2) & (v_2, u_3) \\ (v_3, u_1) & (v_3, u_2) & (v_3, u_3) \end{bmatrix}$$

where (v_i, u_j) is the inner product between the basis vectors from the different bases.

CHAPTER 9   Special Matrices

EXAMPLE 9-12

Consider two different bases for the complex vector space $\mathbb{C}^2$. The first basis is given by the column vectors

$$v_1 = \begin{bmatrix} 1 \\ 0 \end{bmatrix}, \quad v_2 = \begin{bmatrix} 0 \\ 1 \end{bmatrix}$$

A second basis is

$$u_1 = \frac{1}{\sqrt{2}} \begin{bmatrix} 1 \\ 1 \end{bmatrix}, \quad u_2 = \frac{1}{\sqrt{2}} \begin{bmatrix} 1 \\ -1 \end{bmatrix}$$

An arbitrary vector written in the first basis v_i is given by

$$\psi = \begin{bmatrix} \alpha \\ \beta \end{bmatrix}$$

where α, β are arbitrary complex numbers. How is this vector written in the second basis?

SOLUTION 9-12

We construct a change of basis matrix and then apply that to the arbitrary vector ψ. The inner products are

$$(u_1, v_1) = \frac{1}{\sqrt{2}} \begin{bmatrix} 1 & 1 \end{bmatrix} \begin{bmatrix} 1 \\ 0 \end{bmatrix} = \frac{1}{\sqrt{2}}$$

$$(u_1, v_2) = \frac{1}{\sqrt{2}} \begin{bmatrix} 1 & 1 \end{bmatrix} \begin{bmatrix} 0 \\ 1 \end{bmatrix} = \frac{1}{\sqrt{2}}$$

$$(u_2, v_1) = \frac{1}{\sqrt{2}} \begin{bmatrix} 1 & -1 \end{bmatrix} \begin{bmatrix} 1 \\ 0 \end{bmatrix} = \frac{1}{\sqrt{2}}$$

$$(u_2, v_2) = \frac{1}{\sqrt{2}} \begin{bmatrix} 1 & -1 \end{bmatrix} \begin{bmatrix} 0 \\ 1 \end{bmatrix} = -\frac{1}{\sqrt{2}}$$

The change of basis matrix from basis v_i to basis u_i is given by

$$U = \begin{bmatrix} (u_1, v_1) & (u_1, v_2) \\ (u_2, v_1) & (u_2, v_2) \end{bmatrix} = \frac{1}{\sqrt{2}} \begin{bmatrix} 1 & 1 \\ 1 & -1 \end{bmatrix}$$

Applying this matrix to the arbitrary vector, we obtain

$$U\psi = \frac{1}{\sqrt{2}} \begin{bmatrix} 1 & 1 \\ 1 & -1 \end{bmatrix} \begin{bmatrix} \alpha \\ \beta \end{bmatrix} = \frac{1}{\sqrt{2}} \begin{bmatrix} \alpha + \beta \\ \alpha - \beta \end{bmatrix}$$

Quiz

1. Construct symmetric and antisymmetric matrices from

$$A = \begin{bmatrix} -1 & 0 & 2 \\ 4 & 6 & 0 \\ 0 & 0 & 1 \end{bmatrix}$$

2. Is the following matrix antisymmetric?

$$B = \begin{bmatrix} 0 & -1 & 2 \\ -1 & 0 & 6 \\ 2 & 6 & 0 \end{bmatrix}$$

Find its eigenvalues.

3. Is the following matrix Hermitian?

$$A = \begin{bmatrix} 8i & 9 & -i \\ 9 & 4 & 0 \\ i & 0 & 2 \end{bmatrix}$$

4. Show that the following matrix is Hermitian:

$$A = \begin{bmatrix} 2 & 4i & 0 \\ -4i & 6 & 1 \\ 0 & 1 & -2 \end{bmatrix}$$

5. For the matrix in the previous problem, show that its eigenvalues are real.
6. Find the eigenvectors of A in problem 4 and show that they constitute an orthonormal basis.
7. Verify that the rotation matrices R_x, R_y, R_z in three dimensions are orthogonal.
8. Find the eigenvalues and eigenvectors of the rotation matrix R_z.

9.   Is the following matrix unitary?

$$V = \begin{bmatrix} 2i & 7 \\ 1 & 0 \end{bmatrix}$$

10.   Is this matrix unitary? Find its eigenvalues. Are they complex numbers of modulus 1?

$$U = \begin{bmatrix} \exp(-i\pi/8) & 0 \\ 0 & \exp(i\pi/8) \end{bmatrix}$$

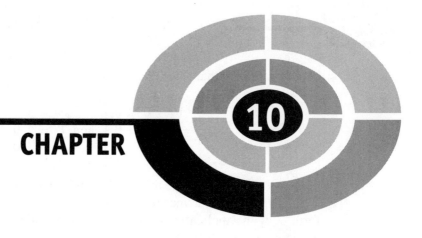

CHAPTER 10

Matrix Decomposition

In this chapter we discuss commonly used matrix *decomposition* schemes. A decomposition is a representation of a given matrix A in terms of a set of other matrices.

LU Decomposition

LU decomposition is a factorization of a matrix A as

$$A = LU$$

where L is a lower triangular matrix and U is an upper triangular matrix. For example, suppose

$$A = \begin{bmatrix} 1 & 2 & -3 \\ -3 & -4 & 13 \\ 2 & 1 & -5 \end{bmatrix}$$

It can be verified that $A = LU$, where

$$L = \begin{bmatrix} 1 & 0 & 0 \\ -3 & 1 & 0 \\ 2 & -\frac{3}{2} & 1 \end{bmatrix} \text{ and } U = \begin{bmatrix} 1 & 2 & -3 \\ 0 & 2 & 4 \\ 0 & 0 & 7 \end{bmatrix}$$

This decomposition of the matrix A is an illustration of an important theorem. If A is a nonsingular matrix that can be transformed into an upper diagonal form U by the application of row addition operations, then there exists a lower triangular matrix L such that $A = LU$.

We recall that row addition operations can be represented by a product of elementary matrices. If n such operations are required, the matrix U is related to the matrix A in the following way:

$$E_n E_{n-1} \cdots E_2 E_1 A = U$$

THE LOWER TRIANGULAR MATRIX *L*

The lower triangular matrix L is found from

$$L = E_1^{-1} E_2^{-1} \cdots E_n^{-1}$$

L will have 1s on the diagonal. The off-diagonal elements are 0s above the diagonal, while the elements below the diagonal are the *multipliers* required to perform Gaussian elimination on the matrix A. The element l_{ij} is equal to the multiplier used to eliminate the (i, j) position.

EXAMPLE 10-1
Find the LU decomposition of the matrix

$$A = \begin{bmatrix} -2 & 1 & -3 \\ 6 & -1 & 8 \\ 8 & 3 & -7 \end{bmatrix}$$

SOLUTION 10-1
Starting in the upper left corner of the matrix, we select $a_{11} = -2$ as the first pivot and seek to eliminate all terms in the column below it. Looking at the matrix, notice that if we take

$$3R_1 + R_2 \to R_2$$

we can eliminate $a_{21} = 6$

$$
\begin{bmatrix} -2 & 1 & -3 \\ 6 & -1 & 8 \\ 8 & 3 & -7 \end{bmatrix} \xrightarrow{3R_1+R_2 \to R_2} \begin{bmatrix} -2 & 1 & -3 \\ 0 & 2 & -1 \\ 8 & 3 & -7 \end{bmatrix}
$$

There is one more term to eliminate below this pivot. The term $a_{31} = 8$ can be eliminated by

$$
4R_1 + R_3 \to R_3
$$

This transforms the matrix as

$$
\begin{bmatrix} -2 & 1 & -3 \\ 0 & 2 & -1 \\ 8 & 3 & -7 \end{bmatrix} \xrightarrow{4R_1+R_3 \to R_3} \begin{bmatrix} -2 & 1 & -3 \\ 0 & 2 & -1 \\ 0 & 7 & -19 \end{bmatrix}
$$

Having eliminated all terms below the first pivot, we move down one row and one column to the right and choose $a_{22} = 2$ as the next pivot. There is a single term below this pivot, $a_{32} = 7$. We can eliminate this term with the operation

$$
-7R_2 + 2R_3 \to R_3
$$

and we obtain

$$
\begin{bmatrix} -2 & 1 & -3 \\ 0 & 2 & -1 \\ 0 & 7 & -19 \end{bmatrix} \xrightarrow{-7R_2+2R_3 \to R_3} \begin{bmatrix} -2 & 1 & -3 \\ 0 & 2 & -1 \\ 0 & 0 & -31 \end{bmatrix}
$$

And so we have

$$
U = \begin{bmatrix} -2 & 1 & -3 \\ 0 & 2 & -1 \\ 0 & 0 & -31 \end{bmatrix}
$$

To find the lower triangular matrix L, we represent each row addition operation that was performed using an elementary matrix. The first operation we performed was

$$
3R_1 + R_2 \to R_2
$$

This can be represented by

$$E_1 = \begin{bmatrix} 1 & 0 & 0 \\ 3 & 1 & 0 \\ 0 & 0 & 1 \end{bmatrix}$$

The second operation was $4R_1 + R_3 \to R_3$. We can represent this with the elementary matrix

$$E_2 = \begin{bmatrix} 1 & 0 & 0 \\ 0 & 1 & 0 \\ 4 & 0 & 1 \end{bmatrix}$$

Finally, we took $-7R_2 + 2R_3 \to R_3$ as the last row addition operation. The elementary matrix that corresponds to this operation is

$$E_3 = \begin{bmatrix} 1 & 0 & 0 \\ 0 & 1 & 0 \\ 0 & -7 & 2 \end{bmatrix}$$

It must be the case that

$$E_3 E_2 E_1 A = U$$

Let's verify this. First we take

$$E_1 A = \begin{bmatrix} 1 & 0 & 0 \\ 3 & 1 & 0 \\ 0 & 0 & 1 \end{bmatrix} \begin{bmatrix} -2 & 1 & -3 \\ 6 & -1 & 8 \\ 8 & 3 & -7 \end{bmatrix} = \begin{bmatrix} -2 & 1 & -3 \\ 0 & 2 & -1 \\ 8 & 3 & -7 \end{bmatrix}$$

Next we have

$$E_2 E_1 A = \begin{bmatrix} 1 & 0 & 0 \\ 0 & 1 & 0 \\ 4 & 0 & 1 \end{bmatrix} \begin{bmatrix} -2 & 1 & -3 \\ 0 & 2 & -1 \\ 8 & 3 & -7 \end{bmatrix} = \begin{bmatrix} -2 & 1 & -3 \\ 0 & 2 & -1 \\ 0 & 7 & -19 \end{bmatrix}$$

and finally

$$E_3 E_2 E_1 A = \begin{bmatrix} 1 & 0 & 0 \\ 0 & 1 & 0 \\ 0 & -7 & 2 \end{bmatrix} \begin{bmatrix} -2 & 1 & -3 \\ 0 & 2 & -1 \\ 0 & 7 & -19 \end{bmatrix} = \begin{bmatrix} -2 & 1 & -3 \\ 0 & 2 & -1 \\ 0 & 0 & -31 \end{bmatrix}$$

To find L, we compute $L = E_1^{-1} E_2^{-1} E_3^{-1}$. The inverses of each of the elementary matrices are easily calculated. These are

$$
E_1^{-1} = \begin{bmatrix} 1 & 0 & 0 \\ -3 & 1 & 0 \\ 0 & 0 & 1 \end{bmatrix}, \quad
E_2^{-1} = \begin{bmatrix} 1 & 0 & 0 \\ 0 & 1 & 0 \\ -4 & 0 & 1 \end{bmatrix}, \quad
E_3^{-1} = \begin{bmatrix} 1 & 0 & 0 \\ 0 & 1 & 0 \\ 0 & \frac{7}{2} & -\frac{1}{2} \end{bmatrix}
$$

and so we obtain

$$
E_2^{-1} E_3^{-1} = \begin{bmatrix} 1 & 0 & 0 \\ 0 & 1 & 0 \\ -4 & 0 & 1 \end{bmatrix} \begin{bmatrix} 1 & 0 & 0 \\ 0 & 1 & 0 \\ 0 & \frac{7}{2} & -\frac{1}{2} \end{bmatrix} = \begin{bmatrix} 1 & 0 & 0 \\ 0 & 1 & 0 \\ -4 & \frac{7}{2} & -\frac{1}{2} \end{bmatrix}
$$

Multiplication by the last matrix gives us the lower triangular matrix

$$
E_1^{-1} E_2^{-1} E_3^{-1} = \begin{bmatrix} 1 & 0 & 0 \\ -3 & 1 & 0 \\ 0 & 0 & 1 \end{bmatrix} \begin{bmatrix} 1 & 0 & 0 \\ 0 & 1 & 0 \\ -4 & \frac{7}{2} & -\frac{1}{2} \end{bmatrix} = \begin{bmatrix} 1 & 0 & 0 \\ -3 & 1 & 0 \\ -4 & \frac{7}{2} & -\frac{1}{2} \end{bmatrix}
$$

Therefore we conclude that

$$
L = \begin{bmatrix} 1 & 0 & 0 \\ -3 & 1 & 0 \\ -4 & \frac{7}{2} & -\frac{1}{2} \end{bmatrix}
$$

Notice that L has 1s along the diagonal. We check that $A = LU$:

$$
LU = \begin{bmatrix} 1 & 0 & 0 \\ -3 & 1 & 0 \\ -4 & \frac{7}{2} & -\frac{1}{2} \end{bmatrix} \begin{bmatrix} -2 & 1 & -3 \\ 0 & 2 & -1 \\ 0 & 0 & -31 \end{bmatrix}
$$

$$
= \begin{bmatrix} (1)(-2) & (1)(1) & (1)(-3) \\ (-3)(-2)+(1)(0) & (-3)(1)+(1)(2) & (-3)(-3)+(1)(-1) \\ (-4)(-2) & (-4)(1)+\left(\frac{7}{2}\right)(2) & (-4)(-3)+\left(\frac{7}{2}\right)(-1)+\left(\frac{1}{2}\right)(-31) \end{bmatrix}
$$

$$
= \begin{bmatrix} -2 & 1 & -3 \\ 6 & -1 & 8 \\ 8 & 3 & -7 \end{bmatrix} = A
$$

Solving a Linear System with an *LU* Factorization

An *LU* factorization allows us to solve a linear system in the following way. Consider the linear system

$$Ax = b$$

Suppose that A is nonsingular. Therefore we can write $A = LU$ and so the linear system takes the form

$$LUx = b$$

Now notice that we can form a second vector using the relationship

$$Ux = y$$

This gives

$$Ly = b$$

Since the matrices U and L are in upper and lower triangular form, respectively, finding a solution is simple because we can use back substitution to solve $Ux = y$ and *forward substitution* to find a solution of $Ly = b$. This is simply carrying out the substitution procedure from top to bottom along the matrix.

FORWARD SUBSTITUTION

Forward substitution works in the following way. Suppose that we had

$$Ly = b$$

$$\Rightarrow \begin{bmatrix} 1 & 0 & 0 \\ -3 & 1 & 0 \\ -4 & \frac{7}{2} & -\frac{1}{2} \end{bmatrix} \begin{bmatrix} y_1 \\ y_2 \\ y_3 \end{bmatrix} = \begin{bmatrix} 2 \\ 12 \\ -2 \end{bmatrix}$$

The first row tells us that

$$y_1 = 2$$

Moving to the next row, substitution of this value gives

$$y_2 = 12 + 3(2) = 12 + 6 = 18$$

Finally, from the last row we obtain

$$y_3 = -2\left[-2 + 4(2) - \frac{7}{2}(18)\right] = -2[-2 + 8 - 63] = 114$$

In short, the solution of $Ax = b$ can be completed using the following steps:

- If A is nonsingular, find the decomposition $A = LU$
- Using forward substitution, solve $Ly = b$
- Using back substitution, solve $Ux = y$ to obtain the solution to the original system x.

EXAMPLE 10-2
Using LU factorization, solve the linear system $Ax = b$, where

$$A = \begin{bmatrix} 3 & -1 & 2 \\ -6 & 3 & 1 \\ 9 & -1 & 1 \end{bmatrix}, \quad b = \begin{bmatrix} 1 \\ 3 \\ 6 \end{bmatrix}$$

SOLUTION 10-2
We use row addition operations to find U. Selecting $a_{11} = 3$ as the first pivot, we eliminate $a_{21} = -6$ with $2R_1 + R_2 \rightarrow R_2$ giving

$$\begin{bmatrix} 3 & -1 & 2 \\ 0 & 1 & 5 \\ 9 & -1 & 1 \end{bmatrix}$$

Next we eliminate $a_{31} = 9$ with $-3R_1 + R_3 \rightarrow R_3$ to obtain

$$\begin{bmatrix} 3 & -1 & 2 \\ 0 & 1 & 5 \\ 0 & 2 & -5 \end{bmatrix}$$

Moving down one row and over to the right one column, we select $a_{22} = 1$ as the next pivot. To eliminate the single term below this pivot, we use the row addition operation $-2R_2 + R_3 \to R_3$, resulting in the upper triangular matrix

$$U = \begin{bmatrix} 3 & -1 & 2 \\ 0 & 1 & 5 \\ 0 & 0 & -15 \end{bmatrix}$$

The elementary matrices that correspond to each of these row operations are

$$2R_1 + R_2 \to R_2 \Rightarrow E_1 = \begin{bmatrix} 1 & 0 & 0 \\ 2 & 1 & 0 \\ 0 & 0 & 1 \end{bmatrix}$$

$$-3R_1 + R_3 \to R_3 \Rightarrow E_2 = \begin{bmatrix} 1 & 0 & 0 \\ 0 & 1 & 0 \\ -3 & 0 & 1 \end{bmatrix}$$

and

$$-2R_2 + R_3 \to R_3 \Rightarrow E_3 = \begin{bmatrix} 1 & 0 & 0 \\ 0 & 1 & 0 \\ 0 & -2 & 1 \end{bmatrix}$$

The inverses of these matrices are

$$E_1^{-1} = \begin{bmatrix} 1 & 0 & 0 \\ -2 & 1 & 0 \\ 0 & 0 & 1 \end{bmatrix}, \quad E_2^{-1} = \begin{bmatrix} 1 & 0 & 0 \\ 0 & 1 & 0 \\ 3 & 0 & 1 \end{bmatrix}, \quad E_3^{-1} = \begin{bmatrix} 1 & 0 & 0 \\ 0 & 1 & 0 \\ 0 & 2 & 1 \end{bmatrix}$$

We have

$$E_2^{-1} E_3^{-1} = \begin{bmatrix} 1 & 0 & 0 \\ 0 & 1 & 0 \\ 3 & 0 & 1 \end{bmatrix} \begin{bmatrix} 1 & 0 & 0 \\ 0 & 1 & 0 \\ 0 & 2 & 1 \end{bmatrix} = \begin{bmatrix} 1 & 0 & 0 \\ 0 & 1 & 0 \\ 3 & 2 & 1 \end{bmatrix}$$

and so

$$L = E_1^{-1} E_2^{-1} E_3^{-1} = \begin{bmatrix} 1 & 0 & 0 \\ -2 & 1 & 0 \\ 0 & 0 & 1 \end{bmatrix} \begin{bmatrix} 1 & 0 & 0 \\ 0 & 1 & 0 \\ 3 & 2 & 1 \end{bmatrix} = \begin{bmatrix} 1 & 0 & 0 \\ -2 & 1 & 0 \\ 3 & 2 & 1 \end{bmatrix}$$

Now we solve the system $Ly = b$. We have

$$\begin{bmatrix} 1 & 0 & 0 \\ -2 & 1 & 0 \\ 3 & 2 & 1 \end{bmatrix} \begin{bmatrix} y_1 \\ y_2 \\ y_3 \end{bmatrix} = \begin{bmatrix} 1 \\ 3 \\ 6 \end{bmatrix}$$

The first row leads to

$$y_1 = 1$$

From the second row, we find

$$y_2 = 3 + 2y_1 = 3 + 2 = 5$$

and from the third row we obtain

$$y_3 = 6 - 3y_1 - 2y_2 = 6 - 3 - 10 = -7$$

Therefore we have

$$y = \begin{bmatrix} 1 \\ 5 \\ -7 \end{bmatrix}$$

To obtain a solution to the original system, we solve $Ux = y$. Earlier we found

$$U = \begin{bmatrix} 3 & -1 & 2 \\ 0 & 1 & 5 \\ 0 & 0 & -15 \end{bmatrix}$$

Therefore the system to be solved is

$$\begin{bmatrix} 3 & -1 & 2 \\ 0 & 1 & 5 \\ 0 & 0 & -15 \end{bmatrix} \begin{bmatrix} x_1 \\ x_2 \\ x_3 \end{bmatrix} = \begin{bmatrix} 1 \\ 5 \\ -7 \end{bmatrix}$$

Using back substitution, from the last line we find

$$x_3 = \frac{7}{15}$$

Inserting this value into the equation represented by the second row, we obtain

$$x_2 = -5x_3 + 5 = 5\left(\frac{7}{15}\right) + 5 = -\frac{7}{3} + \frac{15}{3} = \frac{8}{3}$$

From the first row, we find x_1 to be

$$x_1 = \frac{1}{3}(x_2 - 2x_3 + 1) = \frac{1}{3}\left(\frac{8}{3} - \frac{14}{15} + 1\right) = \frac{1}{3}\left(\frac{41}{15}\right) = \frac{41}{45}$$

SVD Decomposition

Suppose that a matrix A is singular or nearly so. Let A be a real $m \times n$ matrix of rank r, with $m \geq n$. The singular value decomposition of A is

$$A = UDV^T$$

where U is an orthogonal $m \times n$ matrix, D is an $r \times r$ diagonal matrix, and V is an $n \times n$ square orthogonal matrix. From the last chapter we recall that since U and V are orthogonal, then

$$UU^T = VV^T = I$$

That is, the transpose of each matrix is equal to the inverse. The elements along the diagonal of D, which we label σ_i, are called the *singular values* of A. There are r such singular values and they satisfy

$$\sigma_1 \geq \sigma_2 \geq \cdots \geq \sigma_r > 0$$

If the matrix A is square, then we can use the singular value decomposition to find the inverse. The inverse is

$$A^{-1} = \left(UDV^T\right)^{-1} = \left(V^T\right)^{-1} D^{-1} U^{-1} = V D^{-1} U^T$$

since $(AB)^{-1} = B^{-1}A^{-1}$ and $UU^T = VV^T = I$. In the case where A is a square matrix then

$$D = \begin{bmatrix} \sigma_1 & 0 & \cdots & 0 \\ 0 & \sigma_2 & \cdots & \vdots \\ \vdots & \cdots & \ddots & \vdots \\ 0 & 0 & \cdots & \sigma_n \end{bmatrix}$$

Then

$$D^{-1} = \begin{bmatrix} \frac{1}{\sigma_1} & 0 & \cdots & 0 \\ 0 & \frac{1}{\sigma_2} & \cdots & \vdots \\ \vdots & \cdots & \ddots & \vdots \\ 0 & 0 & \cdots & \frac{1}{\sigma_n} \end{bmatrix}$$

If an SVD of a matrix A can be calculated, so can be its inverse. Therefore we can find a solution to a system

$$Ax = b \Rightarrow x = A^{-1}b = V D^{-1} U^T b$$

that would otherwise be unsolvable.

In most cases, you will come across SVD in a numerical application. However, here is a recipe that can be used to calculate the singular value decomposition of a matrix A that can be applied in simple cases:

- Compute a new matrix $W = AA^T$.
- Find the eigenvalues and eigenvectors of W.
- The square roots of each of the eigenvalues of W that are greater than zero are the singular values. These are the diagonal elements of D.
- Normalize the eigenvectors of W that correspond to nonzero eigenvalues of W that are greater than zero. The columns of U are the normalized eigenvectors.
- Now repeat this process by letting $W' = A^T A$. The normalized eigenvectors of this matrix are the columns of V.

EXAMPLE 10-3
Find the singular value decomposition of

$$A = \begin{bmatrix} 0 & -1 \\ -2 & 1 \\ 1 & 0 \end{bmatrix}$$

SOLUTION 10-3
The first step is to write down the transpose of this matrix, which is

$$A^T = \begin{bmatrix} 0 & -2 & 1 \\ -1 & 1 & 0 \end{bmatrix}$$

Now we compute

$$W = AA^T = \begin{bmatrix} 0 & -1 \\ -2 & 1 \\ 1 & 0 \end{bmatrix} \begin{bmatrix} 0 & -2 & 1 \\ -1 & 1 & 0 \end{bmatrix} = \begin{bmatrix} 1 & -1 & 0 \\ -1 & 5 & -2 \\ 0 & -2 & 1 \end{bmatrix}$$

The next step is to find the eigenvalues of W. These are (exercise) $\{0, 1, 6\}$. Only positive eigenvalues are important. The singular values are the square roots, and so

$$\sigma_1 = 1, \ \sigma_2 = \sqrt{6}$$

D is constructed by placing these elements on the diagonal. They are arranged from greatest to lowest in value. There are two singular values, and so D is a 2×2 matrix

$$D = \begin{bmatrix} \sqrt{6} & 0 \\ 0 & 1 \end{bmatrix}$$

Next we find the eigenvectors of W that correspond to the eignvalues $\{1, 6\}$. These must be normalized. We demonstrate with the second eigenvalue. The equation is

$$\begin{bmatrix} 1 & -1 & 0 \\ -1 & 5 & -2 \\ 0 & -2 & 1 \end{bmatrix} \begin{bmatrix} a \\ b \\ c \end{bmatrix} = 6 \begin{bmatrix} a \\ b \\ c \end{bmatrix}$$

This eigenvector equation leads to the relationships (check)

$$b = -5a$$
$$c = 2a$$

Therefore we can write the eigenvector as

$$v = \begin{bmatrix} -a \\ 5a \\ -2a \end{bmatrix}$$

Now we normalize the eigenvector

$$1 = v^T v = \begin{bmatrix} a & -5a & 2a \end{bmatrix} \begin{bmatrix} a \\ -5a \\ 2a \end{bmatrix} = a^2 + 25a^2 + 4a^2 = 30a^2$$

and so we have

$$1 = 30a^2 \Rightarrow a = \frac{1}{\sqrt{30}}$$

And the normalized eigenvector is

$$v = \begin{bmatrix} -\frac{1}{\sqrt{30}} \\ \frac{5}{\sqrt{30}} \\ -\frac{2}{\sqrt{30}} \end{bmatrix}$$

The other normalized eigenvector, corresponding to the eigenvalue $\{1\}$, is (exercise)

$$w = \begin{bmatrix} -\frac{2}{\sqrt{5}} \\ 0 \\ \frac{1}{\sqrt{5}} \end{bmatrix}$$

Now we construct U. The columns of U are the eigenvectors

$$U = \begin{bmatrix} -\frac{1}{\sqrt{30}} & -\frac{2}{\sqrt{5}} \\ \frac{5}{\sqrt{30}} & 0 \\ -\frac{2}{\sqrt{30}} & \frac{1}{\sqrt{5}} \end{bmatrix}$$

Now we compute

$$W' = A^T A = \begin{bmatrix} 0 & -2 & 1 \\ -1 & 1 & 0 \end{bmatrix} \begin{bmatrix} 0 & -1 \\ -2 & 1 \\ 1 & 0 \end{bmatrix} = \begin{bmatrix} 5 & -2 \\ -2 & 2 \end{bmatrix}$$

The eigenvalues of this matrix are $\{1, 6\}$ (why?). The normalized eigvectors are

$$v_1 = \begin{bmatrix} -\frac{2}{\sqrt{5}} \\ \frac{1}{\sqrt{5}} \end{bmatrix}, \quad v_2 = \begin{bmatrix} \frac{1}{\sqrt{5}} \\ \frac{2}{\sqrt{5}} \end{bmatrix}$$

respectively. We construct V by mapping these eigenvectors to the columns of the matrix

$$V = \begin{bmatrix} -\frac{2}{\sqrt{5}} & \frac{1}{\sqrt{5}} \\ \frac{1}{\sqrt{5}} & \frac{2}{\sqrt{5}} \end{bmatrix}$$

(Are these columns orthogonal?). In this case the transpose is equal to V.

We have found the singular value decomposition of the matrix A. Recalling that $A = UDV^T$, we verify the result. First we have

$$DV^T = \begin{bmatrix} \sqrt{6} & 0 \\ 0 & 1 \end{bmatrix} \begin{bmatrix} -\frac{2}{\sqrt{5}} & \frac{1}{\sqrt{5}} \\ \frac{1}{\sqrt{5}} & \frac{2}{\sqrt{5}} \end{bmatrix} = \begin{bmatrix} -\frac{2\sqrt{6}}{\sqrt{5}} & \frac{\sqrt{6}}{\sqrt{5}} \\ \frac{1}{\sqrt{5}} & \frac{2}{\sqrt{5}} \end{bmatrix}$$

and so we obtain

$$UDV^T = \begin{bmatrix} -\frac{1}{\sqrt{30}} & -\frac{2}{\sqrt{5}} \\ \frac{5}{\sqrt{30}} & 0 \\ -\frac{2}{\sqrt{30}} & \frac{1}{\sqrt{5}} \end{bmatrix} \begin{bmatrix} -\frac{2\sqrt{6}}{\sqrt{5}} & \frac{\sqrt{6}}{\sqrt{5}} \\ \frac{1}{\sqrt{5}} & \frac{2}{\sqrt{5}} \end{bmatrix} = \begin{bmatrix} 0 & -1 \\ -2 & 1 \\ 1 & 0 \end{bmatrix} = A$$

QR Decomposition

Let A be a nonsingular $m \times n$ matrix with linearly independent columns. Such a matrix can be written as

$$A = QR$$

The $m \times n$ matrix Q is constructed by setting the columns equal to the orthonormal basis for $R(A)$. R is an $n \times n$ upper triangular matrix that has positive elements along the diagonal. We demonstrate with an example.

EXAMPLE 10-4
Find the QR factorization of

$$A = \begin{bmatrix} 2 & 2 & 0 \\ 0 & 0 & 2 \\ 2 & 1 & 0 \end{bmatrix}$$

SOLUTION 10-4

The column vectors of A are

$$a_1 = \begin{bmatrix} 2 \\ 0 \\ 2 \end{bmatrix}, \quad a_2 = \begin{bmatrix} 2 \\ 0 \\ 1 \end{bmatrix}, \quad a_3 = \begin{bmatrix} 0 \\ 2 \\ 0 \end{bmatrix}$$

To obtain the diagonal elements of R, we compute the magnitude of each vector, using the standard inner product. We find

$$\|a_1\| = \sqrt{4+4} = \sqrt{8} = r_{11}$$
$$\|a_2\| = \sqrt{4+1} = \sqrt{5} = r_{22}$$
$$\|a_3\| = \sqrt{4} = 2 = r_{33}$$

The first column of Q is given by

$$q_1 = \frac{a_1}{\|a_1\|} = \frac{1}{\sqrt{8}} \begin{bmatrix} 2 \\ 0 \\ 2 \end{bmatrix} = \frac{1}{\sqrt{2}} \begin{bmatrix} 1 \\ 0 \\ 1 \end{bmatrix}$$

Next we calculate

$$r_{12} = q_1^T a_2 = \frac{1}{\sqrt{2}} \begin{bmatrix} 1 & 0 & 1 \end{bmatrix} \begin{bmatrix} 2 \\ 0 \\ 1 \end{bmatrix} = \frac{3}{\sqrt{2}}$$

and

$$r_{13} = q_1^T a_3 = \frac{1}{\sqrt{2}} \begin{bmatrix} 1 & 0 & 1 \end{bmatrix} \begin{bmatrix} 0 \\ 2 \\ 0 \end{bmatrix} = 0$$

Using the Gram-Schmidt process, we find

$$v = a_2 - r_{12}q_1 = \begin{bmatrix} 2 \\ 0 \\ 1 \end{bmatrix} - \frac{3}{\sqrt{2}} \left(\frac{1}{\sqrt{2}} \begin{bmatrix} 1 \\ 0 \\ 1 \end{bmatrix} \right) = \begin{bmatrix} 2 \\ 0 \\ 1 \end{bmatrix} - \frac{3}{2} \begin{bmatrix} 1 \\ 0 \\ 1 \end{bmatrix} = \frac{1}{2} \begin{bmatrix} 1 \\ 0 \\ -1 \end{bmatrix}$$

Normalizing gives the next column of Q:

$$q_2 = \frac{1}{\sqrt{2}} \begin{bmatrix} 1 \\ 0 \\ -1 \end{bmatrix}$$

$$r_{23} = q_2^T a_3 = \frac{1}{\sqrt{2}} \begin{bmatrix} 1 & 0 & -1 \end{bmatrix} \begin{bmatrix} 0 \\ 2 \\ 0 \end{bmatrix} = 0$$

The last vector is

$$v = a_3 - r_{13} q_1 - r_{23} q_2 = a_3$$

$$q_3 = \frac{v}{\|v\|}$$

This gives

$$q_3 = \begin{bmatrix} 0 \\ 1 \\ 0 \end{bmatrix}$$

and so we find

$$Q = \begin{bmatrix} \frac{1}{\sqrt{2}} & \frac{1}{\sqrt{2}} & 0 \\ 0 & 0 & 1 \\ \frac{1}{\sqrt{2}} & -\frac{1}{\sqrt{2}} & 0 \end{bmatrix}, \quad R = \begin{bmatrix} \sqrt{8} & \frac{3}{\sqrt{2}} & 0 \\ 0 & \sqrt{5} & 0 \\ 0 & 0 & 2 \end{bmatrix}$$

Quiz

1. Find the LU decomposition of

$$A = \begin{bmatrix} -1 & 2 & 4 \\ 3 & -1 & 1 \\ 2 & 5 & -2 \end{bmatrix}$$

2. Find the LU decomposition of

$$B = \begin{bmatrix} 1 & 2 & 8 \\ 2 & 5 & 1 \\ 3 & 7 & 1 \end{bmatrix}$$

3. Using LU factorization of A, solve the linear system $Ax = b$, where

$$A = \begin{bmatrix} 4 & -1 & 1 \\ 2 & 7 & 0 \\ 1 & 3 & -1 \end{bmatrix}, \quad b = \begin{bmatrix} 1 \\ 1 \\ 1 \end{bmatrix}$$

4. An LDU factorization uses a lower triangular matrix L, a diagonal matrix D, and an upper triangular matrix U to write

$$A = LDU$$

The lower triangular matrix is the same as that used in LU factorization. However, in this case the diagonal matrix D contains the diagonal entries found in the matrix U used in LU factorization. The diagonal elements of U in this case are set to 1. For example, for the matrix

$$A = \begin{bmatrix} 1 & 2 & -3 \\ -3 & -4 & 13 \\ 2 & 1 & -5 \end{bmatrix}$$

we have the LU factorization

$$L = \begin{bmatrix} 1 & 0 & 0 \\ -3 & 1 & 0 \\ 2 & -\frac{3}{2} & 1 \end{bmatrix}, \quad U = \begin{bmatrix} 1 & 2 & -3 \\ 0 & 2 & 4 \\ 0 & 0 & 7 \end{bmatrix}$$

The LDU factorization is

$$L = \begin{bmatrix} 1 & 0 & 0 \\ -3 & 1 & 0 \\ 2 & -\frac{3}{2} & 1 \end{bmatrix}, \quad D = \begin{bmatrix} 1 & 0 & 0 \\ 0 & 2 & 0 \\ 0 & 0 & 7 \end{bmatrix}, \quad U = \begin{bmatrix} 1 & 2 & -3 \\ 0 & 1 & 4 \\ 0 & 0 & 1 \end{bmatrix}$$

Find the LDU factorization of

$$A = \begin{bmatrix} 1 & 11 & -3 \\ 2 & 5 & 4 \\ -3 & 6 & 1 \end{bmatrix}$$

5. Following the SVD example worked out in the text, find the inverse of A by calculating $VD^{-1}U^T$.

6. Find the normalized eigenvectors of

$$\begin{bmatrix} 5 & -2 \\ -2 & 2 \end{bmatrix}$$

7. Find the singular value decomposition of

$$A = \begin{bmatrix} 2 & -1 \\ 0 & 3 \\ 3 & 1 \end{bmatrix}$$

8. Find the QR factorization of

$$A = \begin{bmatrix} 3 & 2 & 0 \\ 8 & -1 & 3 \\ 0 & 4 & 0 \end{bmatrix}$$

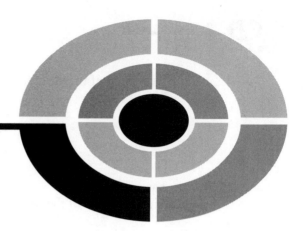

Final Exam

1. For the matrix

$$A = \begin{bmatrix} -1 & 0 & 1 \\ 2 & 4 & 3 \\ 0 & 1 & 0 \end{bmatrix}$$

(a) Calculate $3A$.
(b) Find A^T.
(c) Does A have an inverse? If so, find it.

2. For the matrices

$$A = \begin{bmatrix} 2 & 1 \\ 0 & -1 \end{bmatrix} \text{ and } B = \begin{bmatrix} 3 & -3 \\ 2 & 1 \end{bmatrix}$$

(a) Find $A + B$.
(b) Find $A - B$.
(c) Calculate the commutator of A and B.

3. Determine if the system

$$x + 3y - 7z = 2$$
$$2x + y - 4z = -1$$
$$4x + 8y + z = 5$$

has a solution.

4. Find the trace of the matrix

$$C = \begin{bmatrix} -1 & 4 & 0 & 0 & 2 \\ -5 & 2 & -9 & 0 & 0 \\ 16 & i & 4+2i & 0 & -3 \\ 1 & -5 & 2i & 1 & 0 \\ 0 & -1 & 2 & -1 & 5i \end{bmatrix}$$

5. Find the product of

$$A = \begin{bmatrix} 2i & -7 \end{bmatrix}, \quad B = \begin{bmatrix} 4 \\ 6i \end{bmatrix}$$

6. Prove that if A and B are invertible matrices, then so is their sum, $A + B$.

7. If

$$A = \begin{bmatrix} 1 & -2 & 1 \\ 1 & 1 & -2 \\ -1 & 1 & 2 \end{bmatrix}, \quad A^{-1} = \begin{bmatrix} \frac{2}{3} & x & \frac{1}{2} \\ w & y & z \\ \frac{1}{3} & \frac{1}{6} & \frac{1}{2} \end{bmatrix}$$

what are w, x, y, z?

8. For the matrix A, find x such that A is nonsingular

$$A = \begin{bmatrix} -1 & x & 2 \\ 0 & 1 & 0 \\ 3 & -1 & x \end{bmatrix}$$

9. Solve the system

$$\begin{bmatrix} -2 & 4 \\ 5 & 1 \end{bmatrix} \begin{bmatrix} x \\ y \end{bmatrix} = \begin{bmatrix} -1 \\ 1 \end{bmatrix}$$

by inverting the matrix of coefficients.

10. Find the trace of the commutator of

$$A = \begin{bmatrix} 5 & -1 \\ -6 & 3 \end{bmatrix} \quad \text{and} \quad B = \begin{bmatrix} 4 & 0 \\ 7 & 1 \end{bmatrix}$$

11. Prove that the trace is *cyclic*, i.e., $\text{tr}(AB) = \text{tr}(BA)$. What does this say about noncommuting matrices, if anything?

12. Find the eigenvectors of

$$Y = \begin{bmatrix} 0 & -i \\ i & 0 \end{bmatrix}$$

13. Find a unitary transformation that diagonalizes Y from the previous exercise.

14. Find a unitary transformation that diagonalizes the matrix

$$A = \begin{bmatrix} -1 & 2 \\ 2 & 1 \end{bmatrix}$$

15. Find the eigenvalues of the *Hadamard* matrix

$$H = \frac{1}{\sqrt{2}} \begin{bmatrix} 1 & 1 \\ 1 & -1 \end{bmatrix}$$

16. Find the action of the Hadamard matrix on the vectors

$$v_0 = \begin{bmatrix} 1 \\ 0 \end{bmatrix} \quad \text{and} \quad v_1 = \begin{bmatrix} 0 \\ 1 \end{bmatrix}$$

17. Find the eigenvectors of the Hadamard matrix of problem 15.
18. Find the determinant of

$$A = \begin{bmatrix} -6 & 7 \\ 1 & 5 \end{bmatrix}$$

19. Find the determinant of

$$A = \begin{bmatrix} -1 & 0 & 2 \\ 1 & 4 & 8 \\ 11 & -7 & 6 \end{bmatrix}$$

20. Find the minors of

$$A = \begin{bmatrix} 19 & 5 \\ 2 & -1 \end{bmatrix}$$

21. Calculate the adjugate of

$$A = \begin{bmatrix} 3 & -4 & 1 \\ 0 & 2 & -1 \\ -1 & 6 & 8 \end{bmatrix}$$

22. If possible, find the inverse of

$$A = \begin{bmatrix} -13 & 9 \\ 4 & 11 \end{bmatrix}$$

23. If possible, find the inverse of

$$B = \begin{bmatrix} 6 & 2 & -1 \\ 0 & 1 & 0 \\ 2 & 0 & 2 \end{bmatrix}$$

24. Solve the system

$$2x - 7y = 3$$
$$4x + y = 8$$

using Cramer's rule.

25. Solve the system

$$x + y - z = 2$$
$$2x - y + 3z = 5$$
$$-x + 3y - 2z = -2$$

using Cramer's rule.

26. Find the determinants of

$$1 - t^2, \quad A = \begin{bmatrix} 0 & 1 \\ 1 & 0 \end{bmatrix}, \quad B = \begin{bmatrix} 1 & 0 \\ 0 & -1 \end{bmatrix}$$

where t is a variable (scalar).

27. Find the determinant of the matrix in two ways:

$$B = \begin{bmatrix} 2 & 0 & 0 \\ -3 & 5 & 0 \\ -1 & 2 & 7 \end{bmatrix}$$

28. Is the determinant of this matrix found by the product of the elements on the diagonal the same as the product of its eigenvalues?

$$A = \begin{bmatrix} 8 & -2 & 1 \\ 0 & 4 & -3 \\ 0 & 0 & 1 \end{bmatrix}$$

29. Compute the determinant and trace of the matrix

$$B = \begin{bmatrix} 7 & -9 & 0 & 2 \\ 0 & 1 & 3 & -6 \\ 0 & 0 & 4 & -1 \\ 0 & 0 & 0 & 5 \end{bmatrix}$$

30. Solve the system

$$4x - 3y + 9z = 8$$
$$2x - y = -3$$
$$x + z = -1$$

31. Find the Hermitian conjugate of

$$A = \begin{bmatrix} -i & 4 \\ 3 + 2i & 8 \end{bmatrix}$$

32. Verify the Cayley-Hamilton theorem for

$$B = \begin{bmatrix} 3 & 0 & 1 \\ -2 & 0 & 1 \\ 1 & 4 & 2 \end{bmatrix}$$

33. Construct symmetric and antisymmetric matrices out of

$$A = \begin{bmatrix} 0 & 3 & -2 \\ 1 & 8 & 5 \\ 1 & 4 & -2 \end{bmatrix}$$

34. Find the eigenvalues and eigenvectors for the symmetric and anti-symmetric matrices constructed in the previous problem.
35. Determine if the following matrix is Hermitian or anti-Hermitian:

$$B = \begin{bmatrix} 4 & 2i & 0 & 1 \\ -2i & 3 & 5 & 8 \\ 0 & 5 & 6 & 3-i \\ 1 & 8 & 3+i & -2 \end{bmatrix}$$

36. Find the eigenvalues and eigenvectors of the matrix

$$A = \begin{bmatrix} 4 & 1+i \\ 1-i & -3 \end{bmatrix}$$

37. Is the matrix in the previous problem Hermitian?
38. Prove that the eigenvectors of the matrix in problem 36 constitute an orthonormal basis.
39. Construct a unitary matrix from the eigenvectors of A in problem 36. Use them to transform an arbitrary vector

$$\psi = \begin{bmatrix} \alpha \\ \beta \end{bmatrix}$$

written in the basis

$$v_1 = \begin{bmatrix} 1 \\ 0 \end{bmatrix}, \quad v_2 = \begin{bmatrix} 0 \\ 1 \end{bmatrix}$$

into a vector written in the A basis.

40. Is the following matrix orthogonal?

$$P = \frac{1}{\sqrt{3}} \begin{bmatrix} 1 & -1 & 1 \\ -1 & 1 & 1 \\ 1 & 1 & -1 \end{bmatrix}$$

41. Find the matrix that represents the transformation

$$T(x, y, z) = (2x + y, y + z)$$

from $\mathbb{R}^3 \to \mathbb{R}^2$ with respect to the bases $\{(1, 0, 0), (0, 1, 0), (0, 0, 1)\}$ and $\{(1, 1), (1, -1)\}$.

42. Find the matrix that represents the transformation

$$T(x, y, z) = (4x + y + z, y - z)$$

from $\mathbb{R}^3 \rightarrow \mathbb{R}^2$ with respect to the bases $\{(1, 0, 0), (0, 1, 0), (0, 0, 1)\}$ and $\{(1, 1), (5, 3)\}$.

43. An operator acts on the elements of an orthonormal basis in two dimensions as

$$\sigma v_1 = v_2$$
$$\sigma v_2 = v_1$$

Find the matrix representation of this operator.

44. Describe the transformation from $\mathbb{R}^3 \rightarrow \mathbb{R}^2$ that has the matrix representation

$$T = \begin{bmatrix} 2 & 3 & 7 \\ 1 & -1 & 2 \end{bmatrix}$$

with respect to the standard basis of $\mathbb{R}^3$ and with respect to $\{(1, 1), (1, -1)\}$ for $\mathbb{R}^2$.

45. A transformation from $P_2 \rightarrow P_1$ acts as

$$T(a x^2 + b x + c) = (a - 3b + c)x + (a + b - c)$$

Find the matrix representation of T with respect to the basis

$$\{2x^2 + x + 1, x^2 + 4x + 2, -x^2 + x\}$$

for P_2 and $\{x + 1, x - 1\}$ for P_1.

46. Let $F(x, y, z) = (x + 2z, y - z)$ and $G(x, y, z) = (x + z, y + z)$. Find
    (a) $F + G$
    (b) $4F$
    (c) $-6G$
    (d) $F - 3G$

47. Are the following transformations linear?
    (a) $F(x, y, z) = (x + y, \ y, \ x + y - 4z)$
    (b) $G(x, y, z) = (2x, z)$
    (c) $H(x, y, z) = (xy, yz)$
    (d) $T(x, y, z) = (2 + x, y - z)$

48. An operator acts on a two-dimensional orthonormal basis of $\mathbb{C}^2$ in the following way:

$$Av_1 = 2v_1 + v_2$$
$$Av_2 = 3v_1 - 4v_2$$

   Find the matrix representation of A with respect to this basis.

49. Find the norm of the vector

$$u = \begin{bmatrix} -2 \\ 5+i \\ 4i \\ 1 \end{bmatrix}$$

50. Compute the "distance" between the vectors

$$u = \begin{bmatrix} 2 \\ 4 \end{bmatrix} \quad \text{and} \quad v = \begin{bmatrix} -1 \\ 7 \end{bmatrix}$$

51. Find the inner product of the real vectors

$$a = \begin{bmatrix} 2 \\ -3 \\ 1 \end{bmatrix} \quad \text{and} \quad b = \begin{bmatrix} 1 \\ 5 \\ 1 \end{bmatrix}$$

52. Are the following vectors orthogonal?

$$u = \begin{bmatrix} 1 \\ 0 \\ 1 \end{bmatrix}, \quad v = \begin{bmatrix} -1 \\ 1 \\ 0 \end{bmatrix}$$

53. Are the following vectors orthogonal?

$$u = \begin{bmatrix} 3 \\ -1 \\ 2 \end{bmatrix}, \quad v = \begin{bmatrix} 3 \\ 17 \\ 4 \end{bmatrix}$$

54. Normalize

$$v = \begin{bmatrix} 9 \\ 4 \\ -7 \\ 2 \end{bmatrix}$$

55. What is the distance between

$$a = \begin{bmatrix} 2i \\ 6 \end{bmatrix} \quad \text{and} \quad b = \begin{bmatrix} 6 \\ 2 - 3i \end{bmatrix}$$

56. Construct the conjugate of

$$u = \begin{bmatrix} 2 \\ 3i \\ 5i \end{bmatrix}$$

57. Do the following vectors obey the Cauchy-Schwarz inequality?

$$u = \begin{bmatrix} 1 \\ i \\ 2 \end{bmatrix}, \quad v = \begin{bmatrix} 2 \\ 0 \\ 4i \end{bmatrix}$$

58. Do these vectors satisfy the triangle inequality?

$$u = \begin{bmatrix} 5 \\ 3 \end{bmatrix}, \quad v = \begin{bmatrix} -1 \\ 7 \end{bmatrix}$$

59. Find x so that the following vector is normalized:

$$u = \begin{bmatrix} 3x \\ 8 \\ 2x - 1 \end{bmatrix}$$

60. Find x so that the following vectors are orthonormal:

$$u = \begin{bmatrix} x \\ 7 \\ -1 \end{bmatrix}, \quad v = \frac{1}{\sqrt{2}} \begin{bmatrix} -1 \\ 0 \\ 1 \end{bmatrix}$$

61. If possible, find a parametric solution to the system

$$2x_1 - x_2 + 4x_3 = 8$$
$$x_1 + 3x_2 - x_3 = 2$$

62. Find the row rank of the matrix

$$A = \begin{bmatrix} -2 & 4 & 1 \\ 5 & 0 & 1 \\ 9 & -2 & 11 \end{bmatrix}$$

63. Put the following matrix in echelon form and find the pivots:

$$B = \begin{bmatrix} -6 & 1 & 0 & 1 \\ 1 & 2 & 2 & -3 \\ 3 & 0 & 0 & 2 \end{bmatrix}$$

What is the rank of this matrix?

64. Write down the coefficient and augmented matrices for the system

$$9w + x - 5y + z = 0$$
$$3w - x + 2y - 8z = -2$$
$$4x + z = 12$$

65. What is the elementary matrix that corresponds to the row operation $2R_1 + R_3 \rightarrow R_3$ for a 3×3 matrix?

66. What is the elementary matrix that corresponds to the row operation $R_2 \leftrightarrow R_4$ for a 5×5 matrix?

67. For a 3×3 matrix, write down the elementary matrix that corresponds to $6R_1 - 3R_3 \rightarrow R_3$.

68. Using elementary row operations, bring the matrix

$$A = \begin{bmatrix} -1 & 2 & 4 \\ 5 & 1 & -1 \\ 3 & 2 & -2 \end{bmatrix}$$

into triangular form.

69. For the matrix in problem 68, find the equivalent elementary matrices that correspond to the row operations used.

70. What is the rank of the matrix A in problem 68?

71. Show that the eigenvalues of the matrix A in problem 68 are $(-2, \sqrt{21}, -\sqrt{21})$.

72. Find normalized eigenvectors of the matrix A in problem 68.

73. Are the matrices

$$A = \begin{bmatrix} 6 & -2 & 1 \\ 4 & 0 & 2 \end{bmatrix} \quad \text{and} \quad B = \begin{bmatrix} 1 & -1 & 2 \\ 0 & 0 & 1 \end{bmatrix}$$

row equivalent?

74. If possible, put the following matrix in canonical form using elementary row operations:

$$A = \begin{bmatrix} 1 & 4 & 0 & 2 \\ -2 & 0 & 3 & 8 \\ -5 & 2 & 1 & -2 \\ 6 & 0 & 3 & 1 \end{bmatrix}$$

Identify the pivots.

75. What is the rank of

$$B = \begin{bmatrix} 1 & 2 & 6 \\ 0 & 4 & 1 \\ 0 & 0 & 1 \end{bmatrix}$$

76. Using matrix multiplication, replace row 3 of the following matrix by twice its value:

$$C = \begin{bmatrix} -1 & 0 & 7 & 6 & 1 \\ 0 & 9 & 2 & 3 & 1 \\ 1 & -1 & 5 & 0 & 2 \\ 0 & 0 & 1 & 4 & 8 \\ 5 & 0 & 1 & 0 & 0 \end{bmatrix}$$

77. Determine whether or not the following system has a nonzero solution:

$$4x + 2y - z = 0$$
$$3x - y + 8z = 0$$
$$x + y - 2z = 0$$

78. Use elimination techniques to put the matrix

$$A = \begin{bmatrix} 1 & -2 & 8 & 1 & 4 \\ 2 & -3 & 2 & 2 & 5 \\ 3 & -1 & 1 & 4 & 6 \end{bmatrix}$$

in echelon form.

79. Put the matrix A in problem 78 into row canonical form.

80. Use Gauss-Jordan elimination to put the matrix B in row canonical form where

$$B = \begin{bmatrix} -2 & 1 & 5 \\ 2 & 4 & 1 \\ 3 & 1 & -2 \end{bmatrix}$$

81. Determine if the line $x + 9y = 0$ is a vector space.
82. Show that the set of third-order polynomials $a_3x^3 + a_2x^2 + a_1x + a_o$ constitute a vector space.
83. Explain how to find the row space, column space, and null space of a matrix.
84. By arranging the following set in a matrix and using row reduction techniques, determine if $(2, 2, 3)$, $(-1, 0, 1)$, $(4, -2, 0)$ is linearly independent.
85. Row reduce the matrix

$$A = \begin{bmatrix} 2 & 0 & 1 & 0 \\ -1 & 2 & 0 & 1 \\ 3 & 0 & 1 & 4 \end{bmatrix}$$

86. Find the null space of the matrix A in problem 85.
87. Define a matrix

$$B = \begin{bmatrix} -1 & 2 & 8 \\ 2 & 1 & 1 \\ 3 & 4 & -1 \end{bmatrix}$$

Determine the rank of this matrix and find its eigenvalues.
88. Let

$$B = \begin{bmatrix} -1 & 2 & 5 \\ -2 & 1 & 1 \\ 3 & 4 & -1 \end{bmatrix}$$

Find the row space and column space of this matrix.
89. Write the polynomial

$$v = t^2 + 2t + 3$$

as a linear combination of $p_1 = 2t^2 + 4t - 1$, $p_2 = t^2 - 4t + 2$, $p_3 = t^2 + 3t + 6$.
90. Find the null space of

$$A = \begin{bmatrix} 3 & 2 & 1 \\ 4 & 5 & 6 \\ 6 & 5 & 4 \end{bmatrix}$$

91. Find the row space, column space, and null space of

$$B = \begin{bmatrix} 1 & 3 & 1 & 0 \\ 2 & 1 & 4 & 5 \\ 2 & 7 & 5 & 1 \end{bmatrix}$$

92. Using the inner product $(A, B) = \operatorname{tr}(B^T A)$ for the space of $m \times n$ matrices and using $B = A$ show that it satisfies the properties of a norm.

93. Use the Gram-Schmidt process to find an orthonormal basis for a subspace of the four-dimensional space $\mathbb{R}^4$ spanned by

$$u_1 = \begin{pmatrix} 1 \\ 1 \\ 1 \\ 1 \end{pmatrix}, \quad u_2 = \begin{pmatrix} 1 \\ 2 \\ 4 \\ 5 \end{pmatrix}, \quad u_3 = \begin{pmatrix} 1 \\ -3 \\ -4 \\ -2 \end{pmatrix}$$

94. Calculate the inner product between

$$A = \begin{bmatrix} -1 & 2 & -2 \\ 0 & 1 & 0 \\ 2 & 9 & 1 \end{bmatrix} \quad \text{and} \quad B = \begin{bmatrix} 4 & 1 & 0 \\ 1 & 2 & 3 \\ 4 & 5 & 6 \end{bmatrix}$$

95. Define the difference between orthogonal and orthonormal.

96. Find the eigenvalues and eigenvectors of

$$B = \begin{bmatrix} -1 & 2 & 0 & 1 \\ 1 & 2 & 3 & 4 \\ 0 & 3 & 0 & 1 \\ 2 & 0 & 0 & 2 \end{bmatrix}$$

97. Normalize the eigenvectors of the matrix B in the previous problem.

98. Find the norms of the functions $f = 3x - 4$ and $g = 3x^2 + 2$ on $C[-1, 1]$.

99. Are the functions f and g in the previous problem orthogonal on $C[-1, 1]$?

100. Consider the vector space $\mathbb{R}^3$. Do vectors that have the first component set to zero, i.e., $u = (0, a, b)$ form a subspace of $\mathbb{R}^3$? Do vectors that have the first component set to -1, i.e., $v = (-1, a, b)$ form a subspace of $\mathbb{R}^3$? If not, why not? Here a and b are real numbers.

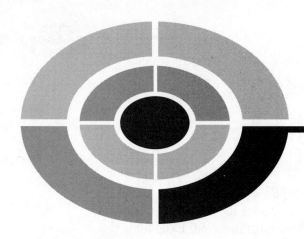

Hints and Solutions to Quiz and Exam Questions

CHAPTER 1

1. Yes that is a solution.
2. $x = 1$, $y = 1$, $z = -1$
3. $x = 13/4$, $y = -3/2$, $z = -5/4$
4. $x = 61/215$, $y = 14/215$, $z = -163/215$

5. $$\begin{pmatrix} 5 & 4 & 1 & | & -19 \\ 3 & 6 & -2 & | & 8 \\ 1 & 0 & 3 & | & 11 \end{pmatrix}$$

6.
$$\begin{pmatrix} 3 & -9 & 5 \\ 3 & 5 & -6 \\ 5 & 0 & 1 \end{pmatrix}$$

7.
$$\begin{pmatrix} 1 & 0 & 0 \\ 0 & 1 & 0 \\ 0 & 2 & 7 \end{pmatrix}$$

8.
$$\begin{pmatrix} 1 & 0 \\ 5 & 3 \end{pmatrix}$$

9. Use
$$\begin{pmatrix} 1 & 0 & 0 \\ 0 & 5 & 0 \\ 0 & 0 & 1 \end{pmatrix}$$

10. Use
$$\begin{pmatrix} 1 & 0 & 0 \\ 0 & 1 & 0 \\ 0 & -2 & 1 \end{pmatrix}$$

CHAPTER 2

1. $A + B = \begin{pmatrix} -1 & 0 & 0 \\ 11 & 8 & 2 \\ 11 & 9 & 1 \end{pmatrix}$, $\alpha A = \begin{pmatrix} -4 & 2 & 0 \\ 18 & 8 & -6 \\ 4 & 2 & 0 \end{pmatrix}$

$$AB = \begin{pmatrix} 0 & 6 & 5 \\ -10 & -17 & 17 \\ 4 & 2 & 5 \end{pmatrix}$$

2. $AB = -1$, $BA = \begin{pmatrix} 2 & -1 & 4 \\ 14 & -7 & 28 \\ 2 & -1 & 4 \end{pmatrix}$

3.
$$AB - BA = \begin{pmatrix} -8 & 5 & 1 \\ -13 & 2 & 10 \\ 17 & 4 & 6 \end{pmatrix}$$

No, because we have

$$AB = \begin{pmatrix} 1-x & 4 \\ x-2 & 4x \end{pmatrix}, \qquad BA = \begin{pmatrix} -x & -1 \\ x+8 & 1+4x \end{pmatrix}$$

No matter what value of x we choose, $(AB)_{12} \neq (BA)_{12}$

5. $\text{Tr}(A) = 16$

7. $\text{Tr}(A) = 2$, $\text{Tr}(B) = 13$

8.
$$A^{\text{T}} = \begin{pmatrix} 1 & 0 & 1 \\ -1 & 4 & 1 \\ 5 & 0 & -2 \end{pmatrix}, \qquad B^{\text{T}} = \begin{pmatrix} 9 & 8 & 16 \\ -1 & 8 & 0 \\ 0 & 4 & 1 \end{pmatrix}$$

9.
$$A^{-1} = \begin{pmatrix} 2/15 & 1/60 & 1/15 \\ -1/15 & 7/60 & 7/15 \\ 0 & 1/4 & 0 \end{pmatrix}$$

10.
$$A^{-1} = \frac{1}{4} \begin{pmatrix} -13 & 9 & -7 \\ -11 & 7 & -5 \\ 21 & -13 & 11 \end{pmatrix}$$

CHAPTER 3

1. $\det|A| = -13$
2. $\det|B| = 324$
3. $\det|A| = -45$, $\det|B| = -26$

$$AB = \begin{pmatrix} -34 & 2 \\ -24 & -33 \end{pmatrix}$$

$$\det|AB| = 1170$$

4. $x = 23/5,\ y = 6/5$
5. $x = 76/33,\ y = 1/3,\ z = -5/11$
6. Follow the procedure used in Example 3-14.
7. Follow Example 3-15:

8. $$\det |A| = -22, \qquad A^{-1} = \tfrac{1}{11}\begin{pmatrix} -4 & 2 & -1 \\ 3 & 4 & -2 \\ -5 & -17 & 7 \end{pmatrix}$$

9. The transpose is

$$\begin{pmatrix} a_{11} & a_{21} \\ a_{12} & a_{22} \end{pmatrix}$$

10. $\det |A| = 24$

CHAPTER 4

1. The sum and difference are

$$v + w = \begin{bmatrix} -1 \\ 12 \end{bmatrix}, \qquad v - w = \begin{bmatrix} -3 \\ -4 \end{bmatrix}$$

2. The scalar multiplication of u gives

$$3u = \begin{bmatrix} 6 \\ -3 \\ 12 \end{bmatrix}$$

3. $a = 2e_1 - 3e_2 + 4e_3$
4. $(u,\ v) = -2 - 12i$
5. $\|a\| = \sqrt{8},\ \ \|b\| = \sqrt{6},\ \ \|c\| = \sqrt{69}$
6. Denoting the normalized vectors with a tilde:

$$a = \frac{1}{\sqrt{14}}\begin{bmatrix} 2 \\ 3 \\ -1 \end{bmatrix}, \qquad u = \frac{1}{\sqrt{19}}\begin{bmatrix} 1+i \\ 4-i \end{bmatrix}$$

7. (a) $u + 2v - w = \begin{pmatrix} 11 \\ 8 \end{pmatrix}$

   (b) $3w = \begin{pmatrix} -3 \\ 3 \end{pmatrix}$

   (c) $-2u + 5v + 7w = \begin{pmatrix} 9 \\ 34 \end{pmatrix}$

   (d) $\|u\| = \sqrt{5}, \quad \|v\| = \sqrt{41}, \quad \|w\| = \sqrt{2}$

   (e) To normalize each vector, divide by the norm given in Part (d).

CHAPTER 5

1. No, does not satisfy closure under addition.
2. Consider the addition of two vectors from this "space,"

$$A = A_x \hat{x} + A_y \hat{y} + 2\hat{z}, \quad B = B_x \hat{x} + B_y \hat{y} + 2\hat{z}$$

$$A + B = (A_x + B_x)\hat{x} + (A_y + B_y)\hat{y} + (2 + 2)\hat{z}$$

$$= (A_x + B_x)\hat{x} + (A_y + B_y)\hat{y} + 4\hat{z}$$

   Since addition produces a vector with z-component $\neq 2$, there is no closure under addition. Therefore this cannot be a vector space.

4. *Hint*: Show that addition and scalar multiplication result in another 2-tuple. Then define the inverse and zero vectors.

5. $$u = (5/4 + i)v_1 + (-3/2 + i)v_2 + (1/4)v_3$$

6. $$v = (39/9)p_1 - 8p_2 - (33/9)p_3$$

7. *Hint*: Show that when you add two such matrices, you get another 2×2 matrix of complex numbers. Also check scalar multiplication and see if you can define a zero vector and additive inverse.
8. Yes (Follow the steps used in Examples 5-11 and 5-12.)
9. Yes (Follow Example 5-16.
10. No
11. *Hint*: Follow the procedure used in Examples 5-18, 5-19, and 5-20.
12. *Hint*: Follow the procedure used in Example 5-21.
13. *Hint*: Follow Examples 5-18, 5-19, 5-20, and 5-21.

CHAPTER 6

1. $(v, u) = v_1^* u_1 + v_2^* u_2 \Rightarrow (v, u)^* = (u, v) = u_1^* v_1 + u_2^* v_2$

2. $(v - 2w, u) = -2 + 16i$

    $2(3iu, v) - (u, iw) = 6i(u, v) - i(u, w) = -3 - i$

3. $(A, B) = 10$
4. *Hint*: Consider the integral of $f^2(x)$ over the interval of interest.
5. The norm is 250/221.
6. No. If $g(x) = -x^3 + 6x^2 - x$, then for the given f we have

$$\int_{-1}^{1} f(x) g(x) \, dx = 848/105 \neq 0.$$

7. No, since $\begin{pmatrix} 1 & 2 & 3 \end{pmatrix} \begin{pmatrix} 0 \\ 2 \\ 5 \end{pmatrix} = (1)(0) + (2)(2) + (3)(5) = 19$

8. Yes. Integrate the product of the functions to show that

$$\int_{0}^{2} f(x) g(x) \, dx = 0$$

CHAPTER 7

1. *Hint*: Check to see if $F(x_1, y_1, z_1) + F(x_2, y_2, z_2) = F(x_1 + x_2, y_1 + y_2, z_1 + z_2)$ and if $\alpha F(x, y, z) = F(\alpha x, \alpha y, \alpha z)$ for some scalar α and repeat for the other transformations.

2.   Try $T = \begin{pmatrix} -3 & 0 & 1 \\ 0 & 2 & 0 \end{pmatrix}$

3.   Try $T = \begin{pmatrix} 4 & 1 & 1 \\ 0 & 1 & -1 \end{pmatrix}$

4. $Z = \begin{pmatrix} 1 & 0 \\ 0 & -1 \end{pmatrix}$

5.   If we let $e_1 = \begin{pmatrix} 1 \\ 0 \\ 0 \end{pmatrix}$, $e_2 = \begin{pmatrix} 0 \\ 1 \\ 0 \end{pmatrix}$, $e_3 = \begin{pmatrix} 0 \\ 0 \\ 1 \end{pmatrix}$

then we find that

$$Te_1 = \begin{pmatrix} 1 \\ 4 \end{pmatrix}, \quad Te_2 = \begin{pmatrix} 2 \\ -1 \end{pmatrix}, \quad Te_3 = \begin{pmatrix} 5 \\ 2 \end{pmatrix}$$

6.   The transformation acts on the standard basis as

$$T(1,\ 0,\ 0) = (2,\ 0,\ 4)$$
$$T(0,\ 1,\ 0) = (1,\ 1,\ -2)$$
$$T(0,\ 0,\ 1) = (1,\ 1,\ -8)$$

7.   You should find that

$$T\left(-x^2 + 3x + 5\right) = x - 2,$$
$$T\left(x^2 - 7x + 1\right) = -5x - 8,$$
$$T\left(x^2 + x\right) = 3x + 1$$

8.

$$F + G = (6x + y + z,\ y - 3z)$$
$$3F = (6x + 3y,\ 3z)$$
$$2G = (8x + 2z,\ 2y - 8z)$$
$$2F - G = (2y - z,\ -y - 2z)$$

9. $A = \begin{pmatrix} 2 & 0 \\ -i & 4 \end{pmatrix}$

10. $T(1,\ 4) = T(2,\ 1) + 4T(1,\ 5) = (1,\ 18,\ 19),$

$T(3,\ 5) = 3T(2,\ 1) + 5T(1,\ 5) = (3,\ 26,\ 36)$

CHAPTER 8

1. The characteristic polynomial is $\lambda^2 - \lambda - 6$.

2. The eigenvalues are $\left(2, \frac{5+\sqrt{13}}{2}, \frac{5-\sqrt{13}}{2}\right)$

4.
$$z_1 = \begin{pmatrix} 1 \\ 0 \end{pmatrix}, \qquad z_2 = \begin{pmatrix} 0 \\ 1 \end{pmatrix}$$

5. Yes
6. The degree of degeneracy is 2 for $\lambda = 4$.
8. To verify that the matrix is unitary, compute the transpose by interchanging rows and columns. Then set $i \to -i$ to construct $U^\dagger$. Finally, show that $UU^\dagger = I$, where I is the identity matrix.
9. Yes
10. The characteristic equation for the matrix is $\lambda^2 - 1 = 0$. Since $X^2 = I$, the matrix satisfies the Cayley-Hamilton theorem.

CHAPTER 9

1. Symmetric and anti-symmetric matrices that can be constructed from A are

$$A_S = \begin{pmatrix} -1 & 2 & 1 \\ 2 & 6 & 0 \\ 1 & 0 & 1 \end{pmatrix}, \qquad A_A = \begin{pmatrix} 0 & -2 & 1 \\ 2 & 0 & 0 \\ -1 & 0 & 0 \end{pmatrix}$$

2. The matrix is symmetric.
3. The matrix is not Hermitian since the conjugate is not equal to A. We have

$$A^\dagger = \begin{pmatrix} -8i & 9 & -i \\ 9 & 4 & 0 \\ i & 0 & 2 \end{pmatrix}$$

4. Take the transpose of the matrix by interchanging rows and columns, and then complex conjugate each element (let $i \to -i$). If you get the same matrix back, it is Hermitian.
5. Use a program like Matlab or Mathematica to find the eigenvalues numerically. They are $(8.54, -2.23, -0.32)$.
6. *Hint*: Normalize each eigenvector. To show they are orthogonal, show that the inner products of the eigenvectors with each other vanish.
7. *Hint*: Show that the inner products among the columns of each matrix vanish.

8. The eigenvalues are $\left(1,\ e^{-i\phi},\ e^{i\phi}\right)$
9. If the matrix were unitary, then $VV.^{\dagger} = I$, where I is the 2×2 identity matrix. This is not true for the given matrix.
10. You should find that $UU^{\dagger} = I$; therefore the matrix is unitary.

CHAPTER 10

1. The LU decomposition is

$$L = \begin{pmatrix} 1 & 0 & 0 \\ -2 & 1 & 0 \\ -4 & 13/5 & 1 \end{pmatrix}, \qquad U = \begin{pmatrix} -1 & 3 & 2 \\ 0 & 5 & 9 \\ 0 & 0 & -87/5 \end{pmatrix}$$

2. The LU decomposition of B is

$$L = \begin{pmatrix} 1 & 0 & 0 \\ 2 & 1 & 0 \\ 8 & -15 & 1 \end{pmatrix}, \qquad U = \begin{pmatrix} 1 & 2 & 3 \\ 0 & 1 & 1 \\ 0 & 0 & -8 \end{pmatrix}$$

3. The LU factorization of A is

$$L = \begin{pmatrix} 1 & 0 & 0 \\ -1/4 & 1 & 0 \\ 1/4 & -1/15 & 1 \end{pmatrix}, \qquad U = \begin{pmatrix} 4 & 2 & 1 \\ 0 & 15/2 & 13/4 \\ 0 & 0 & -31/30 \end{pmatrix}$$

4. $$L = \begin{pmatrix} 1 & 0 & 0 \\ 11 & 1 & 0 \\ -3 & -10/17 & 1 \end{pmatrix}$$

5. The inverse is

$$A^{-1} = \frac{1}{254} \begin{pmatrix} 19 & 29 & -59 \\ 7 & 4 & 5 \\ -27 & 39 & 17 \end{pmatrix}$$

6. The normalized eigenvectors are

$$a_1 = \frac{1}{\sqrt{5}} \begin{pmatrix} -2 \\ 1 \end{pmatrix}, \qquad a_2 = \frac{1}{\sqrt{5}} \begin{pmatrix} 1 \\ 2 \end{pmatrix}$$

7. The singular value decomposition is found numerically to be

$$u = \begin{pmatrix} -0.4 & -0.5 & -0.8 \\ -0.3 & 0.9 & -0.4 \\ -0.9 & -0.1 & 0.5 \end{pmatrix}, \qquad v = \begin{pmatrix} -0.9 & -0.4 \\ -0.4 & 0.9 \end{pmatrix},$$

$$w = \begin{pmatrix} 3.7 & 0 \\ 0 & 3.3 \\ 0 & 0 \end{pmatrix}$$

8. The QR factorization is

$$Q = \begin{pmatrix} 3/\sqrt{73} & 8/\sqrt{73} & 0 \\ 152/\sqrt{111,617} & -57/\sqrt{111,617} & 4\sqrt{73/1529} \\ -32/\sqrt{1529} & 12/\sqrt{1529} & 19/\sqrt{1529} \end{pmatrix}$$

$$R = \begin{pmatrix} \sqrt{73} & -2/\sqrt{73} & 24/\sqrt{73} \\ 0 & \sqrt{1529/73} & -171/\sqrt{111,617} \\ 0 & 0 & 36/\sqrt{1529} \end{pmatrix}$$

FINAL EXAM

1. $3A = \begin{pmatrix} -3 & 0 & 3 \\ 6 & 12 & 9 \\ 0 & 3 & 0 \end{pmatrix}, \qquad A^{\mathrm{T}} = \begin{pmatrix} -1 & 2 & 0 \\ 0 & 4 & 1 \\ 1 & 3 & 0 \end{pmatrix},$

$$A^{-1} = \begin{pmatrix} -2/5 & 1/5 & -4/5 \\ 0 & 0 & 1 \\ 2/5 & 1/5 & -4/5 \end{pmatrix}$$

2. $A + B = \begin{pmatrix} 5 & -2 \\ 2 & 0 \end{pmatrix}, A - B = \begin{pmatrix} -1 & 4 \\ -2 & -2 \end{pmatrix}, [A, B] = \begin{pmatrix} 2 & -11 \\ 6 & -2 \end{pmatrix}$

3. $x = -20/21, y = 23/21, z = 1/21$

4. $\mathrm{Tr}\,(C) = 6 + 7i$

5. $AB = -34i, \qquad BA = \begin{pmatrix} 8i & -28 \\ -12 & -42i \end{pmatrix}$

6. $(A + B)^{-1} = A^{-1} + B^{-1}$

7. $A^{-1} = \begin{pmatrix} 2/3 & 5/6 & 1/2 \\ 0 & 1/2 & 1/2 \\ 1/3 & 1/6 & 1/2 \end{pmatrix}$

8. Looking at the inverse

$$A^{-1} = \begin{pmatrix} \frac{x}{-6-x} & \frac{-2-x^2}{-6-x} & -\frac{2}{-6-x} \\ 0 & 1 & 0 \\ -\frac{3}{-6-x} & \frac{-1+3x}{-6-x} & -\frac{1}{-6-x} \end{pmatrix}$$

So we take $x \neq -6$

9. $x = 5/22,\ y = -3/22$

10. $\operatorname{Tr}(A) = 8,\ \operatorname{Tr}(B) = 5,\ [A, B] = \begin{pmatrix} -7 & 3 \\ -32 & 7 \end{pmatrix}$

11. *Hint*: Write out the summation formula for matrix multiplication.

12. $y_1 = \begin{pmatrix} i \\ 1 \end{pmatrix},\qquad y_2 = \begin{pmatrix} -i \\ 1 \end{pmatrix}$

13. We normalize the eigenvectors of Y and then use them to construct the unitary matrix. It is

$$U = \frac{1}{\sqrt{2}} \begin{pmatrix} i & -i \\ 1 & 1 \end{pmatrix} \Rightarrow U^{\dagger} = \frac{1}{\sqrt{2}} \begin{pmatrix} -i & 1 \\ i & 1 \end{pmatrix}$$

You can verify that $UU^{\dagger} = I$. To diagonalize Y, calculate $U^{\dagger}YU$.

14. *Hint*: The eigenvectors of the matrix are

$$\left\{ \begin{pmatrix} \frac{-1-\sqrt{5}}{2} \\ 1 \end{pmatrix}, \begin{pmatrix} \frac{1+\sqrt{5}}{2} \\ 1 \end{pmatrix} \right\}$$

15. The eigenvalues are $\{-1, 1\}$.

16. $Hv_0 = \frac{1}{\sqrt{2}} \begin{pmatrix} 1 \\ 1 \end{pmatrix},\qquad Hv_1 = \frac{1}{\sqrt{2}} \begin{pmatrix} 1 \\ -1 \end{pmatrix}$

17. $H_1 = \begin{pmatrix} \frac{-2+\sqrt{2}}{\sqrt{2}} \\ 1 \end{pmatrix},\qquad H_2 = \begin{pmatrix} \frac{2+\sqrt{2}}{\sqrt{2}} \\ 1 \end{pmatrix}$

18. $\det|A| = -37$
19. $\det|A| = -182$
20. The minors are
    $(1, 1) \rightarrow 1$
    $(1, 2) \rightarrow 2$
    $(2, 1) \rightarrow 5$
    $(2, 2) \rightarrow 19$
21. Follow Examples 3-13 and 3-14 to find the cofactors of the matrix. Then the adjugate is the matrix of the cofactors.

22. $A^{-1} = \begin{pmatrix} -11/179 & 9/179 \\ 4/179 & 13/179 \end{pmatrix}$

23. $B^{-1} = \begin{pmatrix} 1/7 & -2/7 & 1/14 \\ 0 & 1 & 0 \\ -1/7 & 2/7 & 3/7 \end{pmatrix}$

24. $x = 59/30, \ y = 2/15$
25. $x = 23/11, \ y = 3/11, \ z = 4/11$
26. $\det|A| = \det|B| = -1$
27. $\det|B| = 70$
28. Yes, $\det|A| = 32$, and the eigenvalues are $(1, \ 4, \ 8)$.
29. $\det(B) = 140, \ \text{Tr}(B) = 17$
30. $x = -26/11, \ y = -19/11, \ z = 15/11$.

31. $$A^\dagger = \begin{pmatrix} i & 3 - 2i \\ 4 & 8 \end{pmatrix}$$

32. *Hint*: Find the characteristic equation and insert
$$B^2 = \begin{pmatrix} 10 & 4 & 5 \\ -5 & 4 & 0 \\ -3 & 8 & 9 \end{pmatrix}$$

33. Compute the transpose and then
$$A^{(S)} = \frac{1}{2}\left(A + A^{\mathrm{T}}\right) = \frac{1}{2}\begin{pmatrix} 0 & 4 & -1 \\ 4 & 16 & 9 \\ -1 & 9 & -4 \end{pmatrix}$$

$$A^{(A)} = \frac{1}{2}\left(A - A^{\mathrm{T}}\right) = \frac{1}{2}\begin{pmatrix} 0 & 2 & -3 \\ -2 & 0 & 1 \\ 3 & -1 & 0 \end{pmatrix}$$

34. The eigenvalues of $A^{(A)}$ are $\left(0, -i\sqrt{7/2}, i\sqrt{7/2}\right)$ and the eigenvectors are $a_1 = (1/2, 3/2, 1)$, $a_2 = \frac{1}{10}\left(-2 - 3i\sqrt{14}, -6 + i\sqrt{14}, 10\right)$, $a_3 = \frac{1}{10}\left(-2 + 3i\sqrt{14}, -6 - i\sqrt{14}, 10\right)$.

35. The matrix is Hermitian.

36. The eigenvalues are $\lambda_1 = \frac{1-\sqrt{57}}{2}$, $\lambda_2 = \frac{1+\sqrt{57}}{2}$, and the eigenvectors are $a_1 = \left(1 \quad \left(\frac{7-\sqrt{57}}{4}\right)(1+i)\right)$, $a_2 = \left(1 \quad \left(\frac{7+\sqrt{57}}{4}\right)(1+i)\right)$.

37. Yes the matrix is Hermitian.

38. Show that $(a_1, a_2) = 0$.

39. We can write the matrix as

$$\alpha = \frac{7 - \sqrt{57}}{4}, \qquad \beta = \frac{7 + \sqrt{57}}{4}$$

$$U = \begin{pmatrix} 1 & 1 \\ \alpha(1+i) & \beta(1+i) \end{pmatrix}$$

40. No, check the inner products of the vectors making up the columns.

41. $T = \begin{pmatrix} 1 & 1 & 1/2 \\ 1 & 0 & -1/2 \end{pmatrix}$

42. The action of the transformation on the standard basis is

$$T(1, 0, 0) = (4, 0), \quad T(0, 1, 0) = (1, 1), \quad T(0, 0, 1) = (1, -1)$$

and the matrix representation is found to be

$$T = \begin{pmatrix} -6 & 1 & -4 \\ 2 & 0 & 1 \end{pmatrix}$$

43. $\sigma = \begin{pmatrix} 0 & 1 \\ 1 & 0 \end{pmatrix}$

44. $T = \begin{pmatrix} 3/2 & 1 & 9/2 \\ 1/2 & 2 & 5/2 \end{pmatrix}$

45. Map the transformation onto $\mathbb{R}^3 \to \mathbb{R}^2$ and you should find

$$T\left(2x^2 + x + 1\right) = (0, 2), \qquad T\left(x^2 + 4x + 2\right) = (-9, 3),$$

$$T\left(-x^2 + x\right) = (-4, 0)$$

Then the matrix representation is

$$T = \begin{pmatrix} 1 & -3 & -2 \\ -1 & -6 & -2 \end{pmatrix}$$

46. Using linearity

$$F + G = (2x + 3z), \qquad 4F = (4x + 8z, \ 4y - 4z)$$
$$-6G = (-6x - 6z, \ 6y + 6z), \qquad F - 3G = (2x - z, -2y - 4z)$$

47. H is not linear, but the other transformations are.

48. $A = \begin{pmatrix} 2 & 3 \\ 1 & -4 \end{pmatrix}$

49. $\|u\| = \sqrt{47}$

50. $\|u - v\| = \sqrt{18}$

51. $(a, b) = -12$

52. Yes, $(u, \ v) = 0$.

53. Yes, $(u, \ v) = 0$.

54. The normalized vector is $\tilde{v} = \frac{1}{\sqrt{150}}v$.

55. $\|a - b\| = \sqrt{65}$

56. $u^{\dagger} = \begin{pmatrix} 2 & -3i & -5i \end{pmatrix}$

57. Yes

58. Yes

59. $x = \frac{2 \pm 6i\sqrt{23}}{13}$

60. Notice that v is normalized. In order for the two vectors to be orthogonal, we must have $(u, \ v) = 0$. This leads to the equation $x + 1 = 0$. Setting $x = -1$, we normalize u, giving

$$\tilde{u} = \frac{1}{\sqrt{51}}u$$

Then the set $(\tilde{u}, \ v)$ is orthonormal.

61. First set $x_3 = t$ and then the parametric solution is

$$x_1 = 26/7 - 11/7t$$
$$x_2 = -4/7 + 6/7t$$

62. The rank is 3

63. The rank is 3 and the reduced echelon form is

$$\begin{pmatrix} 1 & 0 & 0 & 2/3 \\ 0 & 1 & 0 & 5 \\ 0 & 0 & 1 & -41/6 \end{pmatrix}$$

64. The augmented matrix is

$$\left(\begin{array}{cccc|c} 9 & 1 & -5 & 1 & 0 \\ 3 & -1 & 2 & -8 & -2 \\ 0 & 4 & 0 & 1 & 12 \end{array} \right)$$

65. The elementary matrix is

$$\begin{pmatrix} 1 & 0 & 0 \\ 0 & 1 & 0 \\ 2 & 0 & 1 \end{pmatrix}$$

66. The elementary matrix is

$$\begin{pmatrix} 1 & 0 & 0 & 0 & 0 \\ 0 & 0 & 0 & 1 & 0 \\ 0 & 0 & 1 & 0 & 0 \\ 0 & 1 & 0 & 0 & 0 \\ 0 & 0 & 0 & 0 & 1 \end{pmatrix}$$

67. The elementary matrix is

$$\begin{pmatrix} 1 & 0 & 0 \\ 0 & 1 & 0 \\ 6 & 0 & -3 \end{pmatrix}$$

68. In triangular form

$$A \rightarrow \begin{pmatrix} -1 & 2 & 4 \\ 0 & 11 & 19 \\ 0 & 0 & -42 \end{pmatrix}$$

69. The elementary matrices are

$$E_1 = \begin{pmatrix} 1 & 0 & 0 \\ 0 & 1 & 0 \\ 3 & 0 & 1 \end{pmatrix}, \quad E_2 = \begin{pmatrix} 1 & 0 & 0 \\ 5 & 1 & 0 \\ 0 & 0 & 1 \end{pmatrix}, \quad E_3 = \begin{pmatrix} 1 & 0 & 0 \\ 0 & 1 & 0 \\ 0 & -8 & 11 \end{pmatrix}$$

70. Rank $(A) = 3$

71. Find the characteristic equation using the determinant to show that the eigenvalues are $\left(-2, \sqrt{21}, -\sqrt{21}\right)$.

72. The eigenvectors are $\{(2, -3, 1), (-2.43, 2.36, 1), (1.23, 1.44, 1)\}$.

73. Two matrices are row equivalent if a series of elementary row operations on one can transform it into the other matrix. Gaussian elimination on A gives

$$\tilde{A} = \begin{pmatrix} 6 & -2 & 1 \\ 0 & 0 & 1 \end{pmatrix}$$

So the matrices are *not* row equivalent

74. Gaussian elimination on A can bring it into the form

$$A \rightarrow \begin{pmatrix} 1 & 4 & 0 & 2 \\ 0 & 8 & 3 & 12 \\ 0 & 0 & -29/4 & -25 \\ 0 & 0 & 0 & -475/29 \end{pmatrix}$$

75. Rank$(B) = 3$

76. Try multiplication by the matrix

$$E = \begin{pmatrix} 1 & 0 & 0 & 0 & 0 \\ 0 & 1 & 0 & 0 & 0 \\ 0 & 0 & 2 & 0 & 0 \\ 0 & 0 & 0 & 1 & 0 \\ 0 & 0 & 0 & 0 & 1 \end{pmatrix}$$

77. Try a parametric solution, set $z = t$, and then the solution is

$$x = -3t/2, \qquad y = 7t/2$$

78. Gaussian elimination gives

$$\tilde{A} = \begin{bmatrix} 1 & -2 & 8 & 1 & 4 \\ 0 & 1 & -14 & 0 & -3 \\ 0 & 0 & 47 & 1 & 9 \end{bmatrix}$$

79. Reduced row echelon form is

$$\begin{pmatrix} 1 & 0 & 0 & 67/47 & 86/47 \\ 0 & 1 & 0 & 14/47 & -15/47 \\ 0 & 0 & 1 & 1/47 & 9/47 \end{pmatrix}$$

80. The reduced row echelon form is the identity matrix.
81. Yes, check scalar multiplication and vector addition.
82. *Hint*: Check linearity.
83. To find the row space, row reduce the matrix A. The row space is made up of the vectors that can be formed from the nonzero rows of the reduced form of the matrix. To find the column space, select the columns in the reduced matrix that have a pivot. These columns are used to form the vectors of the column space. To find the null space, row reduce the matrix; then vectors x that solve $Ax = 0$ and find linear combinations that make up the null space (see Chapter 5).
84. The set is linearly independent.
85. Gaussian elimination can bring the matrix into the form

$$\begin{pmatrix} 2 & 0 & 1 & 0 \\ 0 & 2 & 1/2 & 1 \\ 0 & 0 & -1/2 & 4 \end{pmatrix}$$

86. The null space of the matrix used in Problem 85 is

$$\left\{ \begin{pmatrix} -4 \\ -5/2 \\ 8 \\ 1 \end{pmatrix} \right\}$$

87. Rank$(B) = 3$, numerical evaluation gives the eigenvalues as $(6.01, -5.28, -1.73)$.
88. The row space of B is $\{(1, 0, 0), (0, 1, 0), (0, 0, 1)\}$. Take the transpose of each vector to obtain the column space.
89. $v = (1/83)(72p_1 - 7p_2 + 33p_3)$
90. The null space is

$$\left\{ \begin{pmatrix} 1 \\ -2 \\ 1 \end{pmatrix} \right\}$$

91. The row space is

$$\{(1, 0, 0, 29/17), (0, 1, 0, -13/17), (0, 0, 1, 10/17\}$$

The column space is

$$\left\{ \begin{pmatrix} 1 \\ 0 \\ 0 \end{pmatrix}, \begin{pmatrix} 0 \\ 1 \\ 0 \end{pmatrix}, \begin{pmatrix} 0 \\ 0 \\ 1 \end{pmatrix} \right\}$$

The null space is

$$\left\{ \begin{pmatrix} -29/17 \\ 13/17 \\ -10/17 \\ 1 \end{pmatrix} \right\}$$

92. *Hint*: Show that for matrices of real numbers, $(A, B) = (B, A)$, $(A, A) \geq 0$.

93. *Hint*: Follow the procedure used in Example 6-8.

94. The inner product is $(A, B) = \text{Tr}(B^T A) = 59$.

95. Two vectors u, v are orthogonal if the inner product $(u, v) = 0$. If the vectors are also normalized, i.e. $(u, u) = (v, v) = 1$, then the vectors are orthonormal.

96. This problem should be done numerically with *Matlab* or *Mathematica*. The eigenvalues are $(-1.79 - 0.20i, -1.79 + 0.20i, 1.43, 5.16)$.

97. Compute the eigenvectors numerically. You should find one of them to be

$$\begin{pmatrix} 1.58 \\ 4.36 \\ 2.73 \\ 1.00 \end{pmatrix}$$

98. To find the norms, square each function and integrate over the interval. The norm of f is 38, while the norm of g is 98/5.

99. No they are not because

$$\int_{-1}^{1} (3x - 4)(3x^2 + 2)\, dx = -24 \neq 0$$

100. Vectors with the first component set to zero do form a vector space. To check this, consider vector addition. Vectors with the first component set to -1 do not form a subspace of $\mathbb{R}^3$ because if you add two vectors together, the result no longer belongs to the space of vectors with first component set to -1.

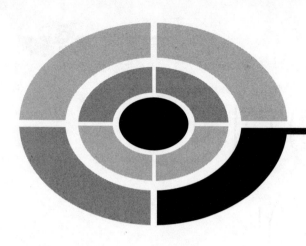

References

Bradley, Gerald, *A Primer of Linear Algebra*. Prentice Hall, New Jersey, 1975.

Bronson, Richard, *Schaum's Outline of Matrix Operations*. McGraw-Hill, New York, 1988.

Lipshutz, Seymour and Lipson, Marc, *Schaum's Oultine of Linear Algebra*, 3rd Edition, McGraw-Hill, New York, 2001.

Meyer, Carl, *Matrix Analysis and Applied Linear Algebra*. Society for Industrial and Applied Mathematics, Philidelphia, 2000.

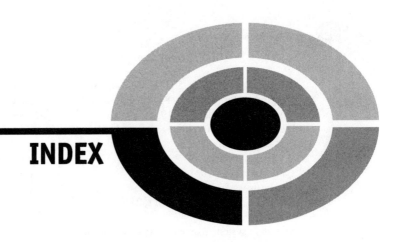

INDEX

Index

Index

About the Author

David McMahon works as a researcher in the national laboratories on nuclear energy. He has advanced degrees in physics and applied mathematics, and has written several titles for McGraw-Hill.